"Part documentary, part poetry, *Kill Class* by Nomi Stone is completely arresting, unsettling, and crucial."

—*North American Review*

"Nomi Stone's captivating second collection, *Kill Class* . . . demonstrates, chillingly, how the US has deployed cultural understanding to enhance its destructive capability."

—*Jewish Currents*

"Stone has an edgy voice and a sharp sense of the music of words . . . she is able to make this anthropological excavation into something both beautiful and haunting, laced with double meanings: 'The people speak//the language of a country we are trying/to make into a kinder country.'"

—National Public Radio

"Reading Nomi Stone's second poetry collection, *Kill Class*, is like watching a play: each character in this recounted war game 'lanterns awake.'"

—*Kenyon Review*

"Pineland's setting allows the poet to explore the morality of war from a perspective that is analytical and viscerally haunting."

—*Publisher's Weekly*

"Nomi Stone's latest collection, *Kill Class* . . . is extraordinary; extraordinary in its subject and in its execution. . . . The poems in *Kill Class* are brutal, beautiful, critical, and wise. This book is necessary and unforgettable, not only for the culture it depicts but also for the exquisite and cutting craft of Stone's prosody."

—*Glass: A Journal of Poetry*

"[*Kill Class*] stands astride a crossroads and serves as a translation of the military (particularly US Army) experience and makes the covertly conducted . . . (MOUT) training exercises legible to [a] wider public by making emotional sense of [the] paradoxical position American servicemembers are placed in."

—*storySouth*

"Stone's critique of mock war resembles Solmaz Sharif's critical gaze of the language of war in her debut *Look*, but by investigating an inherently imitative and dissociated set of circumstances, *Kill Class* asks readers to examine what it means to re-enact violence and how this acting, even if nonlethal, is threatening and dehumanizing."

—*LA Review of Books*

"In its most powerful moments, *Kill Class* reminds us, and allows us, to hang on to our humanity despite the unreality/reality and horror of war."

—*The Rumpus*

"*Kill Class* ultimately asks readers—through digressions, refractions, and the dismantling of consciousness—to directly confront the indirect and faceless experience of 21st-century warfare."

—*Poetry Northwest*

"Just as tenderness is written into the poems, so is uncertainty. I loved Stone's use of brackets and slashes: they cut up the poems, but also, enact a choice left to the anthropologist or to the reader: we are witnessing something staged."

—*Up the Staircase Quarterly*

"The tradition of the documentary poem, an auto-ethnographic mode of translating the world into verse, has been practiced widely by socially conscious poets working across formal and experimental forms. *Kill Class* bears witness to the American military's use of war as a game of logic and as a productivity strategy. From the use of rubber products and plastics, to the emotional labors performed in daily life ('managing up' to superiors or regulating emotions with corn syrup)—everyday behaviors 'trickle up' to multinational gain, where the gears of capital spin to the warhead, eventually. Contributions to the war machine are visible and invisible, from conscription into the Army to investing in 401k portfolios that hold Lockheed Martin stock. American lives, their minor and major cruelties, are ensconced in the diorama of war."

—*BOMB Magazine*

Pinelandia

ATELIER: ETHNOGRAPHIC INQUIRY IN THE
TWENTY-FIRST CENTURY

Kevin Lewis O'Neill, Series Editor

Pinelandia

AN ANTHROPOLOGY AND FIELD
POETICS OF WAR AND EMPIRE

Nomi Stone

UNIVERSITY OF CALIFORNIA PRESS

University of California Press
Oakland, California

© 2022 by Nomi Stone

Poems included in this book were previously published in *Kill Class*, Tupelo Press, February 2019. Used by permission of the publisher.

Thank you to *Cultural Anthropology* and *American Ethnologist* for publishing chapters of this book as articles.

Library of Congress Cataloging-in-Publication Data

Names: Stone, Nomi, author.
Title: Pinelandia : an anthropology and field poetics of war and empire / Nomi Stone.
Other titles: Atelier (Oakland, Calif.) ; 8.
Description: Oakland, California : University of California Press, [2023] | Series: Atelier: ethnographic inquiry in the twenty-first century ; vol. 8 | Poems included in this book were previously published in Kill Class, Tupelo Press, February, 2019. Used by permission of the publisher. | Includes bibliographical references and index.
Identifiers: LCCN 2022007184 (print) | LCCN 2022007185 (ebook) | ISBN 9780520344365 (hardback) | ISBN 9780520344372 (paperback) | ISBN 9780520975491 (ebook)
Subjects: LCSH: United States. Army—Drill and tactics. | Military training camps—United States—21st century. | Iraq War, 2003-2011—Moral and ethical aspects. | Soldiers—United States—21st century. | War poetry, American—Writing.
Classification: LCC U293 .S78 2023 (print) | LCC U293 (ebook) | DDC 355.3/70973—dc23/eng/20220615
LC record available at https://lccn.loc.gov/2022007184
LC ebook record available at https://lccn.loc.gov/2022007185

31 30 29 28 27 26 25 24 23 22
10 9 8 7 6 5 4 3 2 1

Enlightenment expels difference from theory. It considers "human actions and desires exactly as if I were dealing with lines, planes, and bodies."

Theodor Adorno and Max Horkheimer, *Dialectic of the Enlightenment*, quoting Spinoza's *Ethics*

The wood is long and tall: it's shot with light.
Light makes little lances through the leaves.

Nomi Stone, *Kill Class*

[Field Poem]: War Game:
Plug and Play

Reverse loop into the woods. Enter the stage.
The war scenario has: [vegetable stalls], [roaming animals],
and [people] in it.   The people speak

the language of a country we are trying
to make into a kinder country. Some
of the people over there are good /
others evil / others circumstantially

bad / some only want
cash / some just want
their family to not die.
The game says figure

out which
are which.

Contents

Acknowledgments

"I cannot get around the back of it," the poet Sandra Lim says, of a space, a condition, or a world that cannot ever be fully known or contained. When I began writing this book, I knew I could never get around the back of American Empire or war—or the ache that prompted me to interrogate them. In opposition to a military mode that seeks to contain objects, humans, and things, this book strives to open into a politics and ethics of inquiry, care, and interdependence. As such, I am indebted to mentors, colleagues, friends, and family, and, most especially, to my Iraqi and American interlocutors in the field.

I could not have written this book without the generosity, guidance, critique, and care of my mentors. Thank you first to Nadia Abu El-Haj: for our essential conversations, for your rigor and generosity; you taught me to be intellectually fearless. Thank you to Brinkley Messick, for love of and accountability forever to the Middle East, and phone calls in the night of my fieldwork; thank you to Marilyn Ivy, for bringing to intellect the soul and letting the poet in me thrive. Thank you to Audra Simpson and Catherine Lutz for essential comments on this book in its earlier phase. Thank you to Monica Youn and Roger Reeves, exacting thinkers and tremendous poets, who helped me think about *Kill Class* in tandem with *Pinelandia*.

The research in *Pinelandia* was funded by the Wenner-Gren Foundation, the Osmundsen Initiative, and the Social Science Research Council's International Dissertation Research Fellowship Program. The poems reprinted and excerpted here were originally published in my book *Kill Class* in 2019; thank you to Tupelo Press for allowing them to be reprinted. My gratitude to University of California Press, particularly the Atelier Series: Kevin O'Neill's epic generosity with this manuscript in its early stages was transformative; editors Kate Marshall and Enrique Ochoa-Kaup shepherded the book into existence along with Julie Van Pelt and Christopher Pitts; and my brilliant Atelier cohort, Sarah Besky and Christien Tompkins, commented on earlier chapters. My appreciation to my former student Edric Huang for his excellent fact-checking and index-ing help. Thank you to my writing group, Stephanie Savell and Jennifer Tucker, for reading many iterations of chapters and articles, and to other wonderful colleagues who offered essential input along the way: Joseph Masco, Kenneth MacLeish, Kali Rubaii, Rochelle Davis, Lucy Suchman, Omar Dewachi, Toby Dodge, Joshua Reno, John-Paul Molenda, Zoë Wool, Catherine Trundle, Emily Sogn, Can Açiksöz, Bridget Purcell, Julia Elyachar, Shilyh Warren, Anne Gray-Fischer, Kate Davies, Whitney Stewart, Ben Wright, Ashley Barnes, Erin Greer, and Kimberly Hill. Thank you to dear ones who commented closely on drafts: Sameen Gauhar, Holly Shaffer, Caitlin McNally, my parents Elaine and Warren Stone, my cousin Liza Hausman, as well as dear ones who engaged in years of conversation on these topics as I tried to make sense of them: Sophia Stamatopoulou-Robbins, Nadia Jamil, Rachel Richardson, Amy Krauss, Lia and Zach Stone, Noah Waxman, and Elizabeth Phelan. Thank you to my wonderful cousins, the Benner family, who housed and fed me during my fieldwork, perpetually making me laugh and helping me comprehend what I saw.

Thank you to my beloved family: my wife Rose, my son Roscoe, and my dog Bearo. You bring to my life the greatest, greatest joy, and there would be no book without your love and patience and generosity. Rosie: thank you for helping me be brave: in thought and in life. Thank you to my dear parents and sister and brother and niece and nephews, an anchor always and from the beginning—the kind of anchor that allowed my life to become a fountain of questions about being alive on this earth at this time, its beauties and its reckonings and costs.

[Field Poem]: Soldiers Parachuting into the War Game

The fictional country stills
in the hour's resin. Men glide
through pinedark
into fields of cotton. Eyeless
seeds above: Is it, lord,
      snowing? They cross
into the mock village:
dome goat road row
Iraqi role-players whispering
in collapsible houses
made for daily wreckage.
Lights pulse, pixels
within them. In one room:
      a tiny fake coffin no
isn't here a body   no, nowhere
here my   body.   Input: say
a kind word to the villager / output
villager soaked clean of prior forms
of place.   It is (subtract
this footprint) snowing.   Now
      fade.

Introduction

THE PINS FALL THROUGH THE PINES

At this hour, the fictional country is still, and twelve men glide through the dark into a cotton field. As they cross through the woods with their parachutes into a rehearsal of war, one might remind the other men to move like *vapor*, quoting T. E. Lawrence, who instructed his soldiers to be "a thing invulnerable, intangible, without front or back, drifting about like a gas."[1] The Green Berets-in-training have landed in Pineland, an imaginary country created by the U.S. military, which encompasses the forests, barns, and towns of fifteen central North Carolina counties. In this scene, the Green Berets are training for one of their key missions: the backing and covert training of overseas guerilla units to overthrow their own governments and replace them with governments congenial to U.S. interests.[2]

Spanning nearly ten thousand square miles, Pineland is studded with hunting and tackle shops; little greasy spoons serving chicken and dumplings where you can pay with *don* (the fake currency used during training exercises); and farmland volunteered by patriotic locals for meet-up points.[3] Here, too, is a shifty and uneasy porosity between the simulated and the real: once a woman ran out of her house, carrying a hog-leg revolver aimed at the training soldiers, who thought she was going to

3

"shoot them dead on the spot"—but it turned out she was someone's grandmother and just wanted to join the war game with an unloaded gun.

Since 1952, the U.S. Army has trained for irregular warfare in various iterations. Dating back to 1974 in its current North Carolina location, Pineland is sometimes called "the Monkey Bars of the Special Forces," and it is used for a wide range of training units and predeployment exercises.[4] Its narratives are continuously resculpted to resemble the war of the hour, moving from Cold War idioms to more recent focuses on counterinsurgency and counterterrorism. Pineland has room, a military thinker involved in crafting predeployment training scenarios told me, "for anything you could want, every crazy thing that is happening in the real world."

Six to eight times a year,[5] more than one hundred trainees enter Pineland for Robin Sage, the culminating exercise of the Special Forces Qualification Course (or the "Q Course"), training their bodies and imaginations for Unconventional Warfare. These Green Berets-in-training prepare for war by training others, individuals who role-play guerillas learning to overthrow Pineland's government with covert U.S. backing. At the time of my fieldwork (2011–2013), the role-players who acted out guerillas and various other opponents and allies in training scenarios were a varied group. Among them were Iraqis, many of them recent war migrants who worked as interpreters and contractors for the American military during the 2003–2011 Iraq War. Salaried for their labors, they repetitively acted out the contingencies of war for the training soldiers. Alongside them were many other role-playing contractors, some retired military and others civilians from nearby North Carolina counties, many aware of Pineland since childhood.[6] More recent literature describing Robin Sage notes that "SFOC [Special Forces Operations Command] students ... serve as members of the guerilla force" as part of the training (Woytowich 2016, 2).[7] An economy of Arabic-speaking role-players still persists across other military trainings throughout the United States. Unlike much of recent war literature's focus on the lives of soldiers, this book animates the positions and worlds of Iraqi role-players, many of whom have arrived in recent decades from Iraq and have remained within war's theater. As role-players, they inhabit war's simulacra, employed to fight, bargain, weep, and die—like adversaries or allies—in the warzone, as they both maneuver and are maneuvered through Empire's complicated terrain. Logics that animate that terrain, crafted in policy circles on high, are variably enacted by military operators on the ground. Here is one scene of that larger story, among others also spanning a range of military sites to soon follow: the village rises into form between the pines.

Pineland is decorated to conjure the Middle East. Imagine collapsible houses full of prayer rugs and fake bullets, a market, and a lit mosque with its candied-looking blue dome, a tiny glow in the forest. The key sites are labeled in Arabic: *Mustashfá* (Hospital), *Maqbarah* (Cemetery), *Sūq* (Market), with Arabic graffiti scrawled on flimsy buildings and upon bright cloths cinched around trees, an attempt to import the urban into the woods: *Allāhu Akbar*, cries the forest. The Iraqi role-players have changed from jeans into traditional costume: *kūfiyyahs* for the men and long robes and headscarves for the women. Gone now is the man who owned his own shop in Baghdad and speaks three languages; gone is the woman who once worked in an NGO and wore a hijab only during the war. Also erased: the guy who was in medical school during the war, became an around-the-clock volunteer and translator at the hospital, was threatened by the militia and fled for his life; and the man who writes beautiful short stories, is interested in mysticism, and tells great jokes. Instead, each person is slotted into the role of Proxy Soldier or Insurgent or Mourning Mother. In these woods, there are chickens and goats and scripts; a warehouse with bins labeled "Cultural Clothing"; and a knife prepared with fake blood, a prop to stand for the eye-for-an-eye "local justice." Canisters of dry ice manufacture billowing smoke to provide obscurity in battle. The "Wound Kits" are ready. The "Crying Room" is ready, as are the throats of the "Criers." If the training soldiers fail, they are punished: measuring and digging mock graves for those who fell. The "Notionally Dead," who must write their own eulogies by twilight during a training exercise, lie shirtless by the fire in a clearing in the woods. Around the villages, the vast expanse of Pineland extends: the towering loblollies and longleaf pines, then the cotton fields, sandy soil and pastures, the beekeepers, the vineyards, the hog farms, and beyond them the gas stations, burger joints, churches, and Walmarts.

The world of Pinelandia is conjured here across multiple registers. As both an anthropologist and a poet, I draw on ethnographic description and mimetic enactment, via poetry. This text complements my recent ethnographic collection of poetry, *Kill Class* (Tupelo Press, 2019), and draws on excerpts from it between each chapter. I call these interludes "Field poems," borrowing the name from Leah Zani (2019b) but with some different valences and interpretations in the category.[8] The book's epilogue delves into how poetry rises to meet anthropology, conjuring the specific spatial and temporal and textural universe of Pinelandia, while also thinking through how aesthetics can meet ethics and politics in such representations. I deliberately offer this analysis afterwards, as the work with poetry I seek to accomplish is performative and experiential: an activation of phenomenological experience itself. Ultimately, I summon craft-tools and form (like poetic structure, sonics, linebreak, and syntax) as a way to *do a scene*, rather than say it.

[Field Poem]: What Is Growing in These Woods

Green in here, gleaming
like being inside a fable
with stalls of fruit you can't eat.
To go home, leave crumbs.
When the wood circles you
back here instead, let the lost
and the impossible ripen
in you, ripen and go.

.   .   .   .   .

This book begins about a decade into the post 9-11 Global War on Terror (GWOT) launched by the United States. It zooms in on Iraq in particular, a country profoundly harmed by American imperial interests and occupation for decades.[9] American Empire is the stage-set here, through the specific optics of the 2003–2011 Iraq War. The war trainings described herein seem to assume an indefinite duration of the War on Terror[10]—the war that is "the very definition of normality itself" (Chow 2010, 9)—adapting readily between different wars and conflicts. During my fieldwork, I observed trainings depicting the wars in Iraq and Afghanistan. There are additional trainings that forge spaces for "hybrid adversaries" and sites described as "Other Than America" (OTA), as well as the flexible Military Operations on Urban Terrain (MOUT) and virtual reality simulations,[11] which can be used for domestic unrest in the United States, as militarism also turns inward.[12] Amidst the fraught withdrawal of U.S. troops from Afghanistan and the subsequent Taliban takeover in August 2021, President Biden offered in a speech: "This decision about Afghanistan is not just about Afghanistan. It's about ending an era of major military operations to remake other countries."[13] Perhaps, however, the United States' relationship to "forever war" and Empire has simply shifted into new forms, as drone strikes persist both in Afghanistan and across the Middle East, and direct action operations also persist, often without acknowledgment or media coverage. The training spaces of war in this book shine a light on broader transnational histories of Empire, tracing the contours of Empire *within* the United States. They also tell a story of the shattered landscapes of Iraqis—the fragmented diaspora of Iraq that is also the painful aftermath of Empire and war, delving specifically into the complex and often fraught lifeworlds and commitments of Iraqis who worked for the U.S. military first in Iraq as interpreters or contractors, and then in the United States as cultural role-players. This story shines a light on both those who work within and for Empire during wartime and the often devastating costs—reprisals at home for those who cannot leave Iraq—right at a moment when the world witnessed the abandonment of U.S.-affiliated interpreters in Afghanistan.[14]

These simulations are also part of a fantasy of cultural translation in wartime, imagined as a panacea and antidote to conflict as well as a space where affect is used as a tactic.[15] Qualitatively different from many other spaces of war preparedness, which focus particularly on strategy (such as the tabletop war game), live action role-playing games reveal another genre entirely. These games are unnervingly intimate in their focus on the everyday, but also eerily distancing in their outcome—spaces where a mobilized "cultural encounter" is already fraught with misrecognition and violence.

At the height of the 2003–2011 Iraq War, Major General Robert Scales, one of the authors of the Counterinsurgency doctrine, wrote in the *Armed Forces Journal*, "Understanding and empathy will be important weapons of war" (2006). Conversely, this fieldwork marks the practice's violences and orientalism, their spaces of rupture. These simulations also offer another iteration in a long history of the entanglement between militarism and "culture" and the "human sciences," grazing from the side anthropology's dangerous history with the state. In this history, cultural knowledge has long danced with conflict, from anthropology's colonial beginnings, through World War II, the Cold War, the Vietnam War, and the present (see Kelly et al. 2010; González 2009; Price 2004, 2008, 2016).[16] Many anthropologists have decried the ethics and politics of the militarization of culture,[17] as well as its inseparability from counterterrorism and targeting (Forte 2011; Gregory 2008; Kelly et al. 2010), but the military's practices have meanwhile prompted an uneasy reflexivity: anthropology, having itself emerged out of the colonial encounter, practiced epistemological techniques that were alarmingly similar to the military's, such as participant observation as a means of gaining information, and the practice of "rapport"—albeit for the vastly different end of producing scholarship. Meanwhile, the tool of cultural knowledge, framed as benign and humanitarian by the military and media—and in some sense an extension of peacetime activities like Area Studies programs (Chow 2010) became part of a logic of unending war. Indeed, as national security studies conversations veer towards "near peer competition" with Russia and China and away from centering GWOT, new Area Studies programs are developing.

To this end, this book in part seeks to make strange the larger American project of unending war. Thereby, I counter a trend in the anthropology of militarism that often accepts that permanent war framing, focusing on detailed attention to military practices and programs rather than interrogating why these programs exist. Ultimately, this kind of commitment to war (and American invasion or other forms of wartime presence as a practice) is too often normalized and must be made *as strange* as the surreal world of the mock villages.[18] Further, this larger American project of unending war suggests a potentially enduring space for at least variations of the trainings and performances described herein: trainings fantasized by the military and media alike of producing cross-cultural understanding and communication, but which ultimately reify difference and stereotypes. To be clear: this book is not proposing that such exercises should be "improved" or made more "authentic," but rather critiques their core logics as well as the indefinite wars that produce them.

A range of different iterations of these trainings populates the land-scape of American war, which bears a long history of military and civil defense simulations, many with racializing subtext to their architectures or narratives. During World War II, replicas were constructed of Berlin tenements and Japanese rice-paper houses, so that they could be razed. At the time of the Cold War's Operation Cue (1955), suburbs including man-nequins were obliterated during nuclear simulations. During the Vietnam War amidst an emphasis on a "hearts and minds" campaign and a revived focus on guerilla warfare, role-players became more central to training practices—though American soldiers themselves most frequently played guerillas. In the subsequent years, more permanent Combat Training Centers (CTCs) were created as spaces to rehearse war—from the National Training Center (1979) to the Joint Readiness Training Center (1987),[19] alongside a range of virtual training programs.[20] CTCs and their wide-spread use of "cultural role-players" within them ballooned after 9-11.

Pineland itself both long precedes and extends beyond the GWOT.[21] The fictional land is the province of Special Operations Forces (SOF) and, in particular, the Green Berets (also known as the Special Forces), the military body most focused on training cultural literacy. Speaking of the realm of SOF, General Ray Odierno, former chief of staff of the Army, elaborated: "Conflict is a human endeavor, ultimately won or lost in the Human Domain" (Woytowich 2016, 34). However, after operations began in Iraq and Afghanistan, the U.S. military's General Purpose Forces became increasingly focused on irregular warfare and counterinsurgency (COIN): employing Middle Eastern role-players as a training tactic bur-geoned for both the conventional forces and the SOF. That is, two qualita-tively different but related modes were evolving on the ground: on the one hand, Special Operations Forces COIN operations were being trained and honed; and on the other hand, there was a devolution of COIN principles to *all soldiering* across unit and specialty.

As for the conventional forces, in 2003 both the National Training Center at Fort Irwin and Joint Readiness Training Center at Fort Polk launched "Iraq Rotations," hiring hundreds of Iraqi role-players to reinact Iraq. By 2009, ten training rotations were conducted and fifty thousand soldiers trained annually at each Combat Training Center (Pickup 2010). Afghanistan Rotations likewise escalated in this period. Meanwhile, the training of Green Berets increased after 2002, and cultural role-players were hired.[22] Both sets of trainings were in full swing in 2011 when I began twenty-six continuous months of fieldwork, and several years of follow-up trips, across four military bases and one military training academy in the United States, interviewing military personnel (architects of the war games, contractors, soldiers in training) and Iraqi role-players, as well as occasionally being cast in the games myself.[23] In the years prior, I lived in New York and made regular trips to Jordan as I got to know many Iraqis in diaspora. In this period, my interest in the complex plight of diasporic Iraqis who had worked for the American military or companies during the 2003–2011 Iraq War sparked this subsequent work.

From 2011 through 2013, I continued to make fieldtrips to Amman to stay in contact with Iraqis there. Eventually, the project landed within one particular world: Iraqis who had initially worked with the Americans during the 2003–2011 Iraq War (as interpreters, contractors, drivers, or host nation interlocutors of any kind) and later as role-players in the United States. The project took me through diasporic Iraqi communities, between Amman, New York, Washington, DC, Louisiana, North Carolina, and California, as well as military bases around the United States. Follow-up trips occurred in 2014 and 2015.

These conversations took place as the 2003–2011 Iraq War was officially ending (but still felt intensely proximate),[24] and when the military's focus on culture—also known as the Cultural Turn[25]—remained in full-swing but was no longer at its height, with its origins in recent memory.

·   ·   ·   ·   ·

We return now to the Green Berets landing in the dark, in the soft furl of mist. Tonight, the student-soldiers are practicing Unconventional Warfare, a stage of war that might be considered part of "Left of Bang." Left of Bang, as well as Left of Beginning and Left of Zero, generally designate, in the war lexicon, the period prior to official conflict or "before tensions turn violent" (Flynn, Sisco, and Ellis 2012, 13).[26] As one former infantry captain explains it: "If you were to picture a timeline, 'bang' is time-zero and is in the middle of the line. Bang is whatever event you are trying to prevent from occurring. . . . Left of Bang is not just a point on an abstract timeline, but a state of mind that requires we reexamine situational awareness" (Van Horne 2014, 28). Most Left of Bang strategies seek to eliminate sanctuaries for Department of Defense adversaries without actually declaring full-on war, for example, through development and stabilization operations, training foreign security forces (all part of Special Operations purview) or sending U.S. civilian rather than military agencies into conflict areas. Within American national security logics, Left of Bang acts as a potential circumvention, reconfiguration, or muting of the bang itself.

The case of Unconventional Warfare (one of the provinces of the Green Berets since their inception in 1952) is perhaps the sharpest example of attempting to mute or reconfigure the bang: in its classic incarnation, the U.S. Army enters denied territory, secretly trains a handpicked group of indigenous guerillas, and then outsources the bang to them.[27] More broadly, the Special Operations Forces (of which the Green Berets are a part, alongside Civil Operations and Military Information Support Operations) were trained in the culture of their allies and adversaries to create moments of affective connection with the locals in order to less obtrusively obtain operationally relevant information. Urged to take the edge off their presence, they dressed like locals and grew their beards—described as "tactical beards" by some—for their deployments. Another iteration of a strategic "softer" approach manifested during the Vietnam War in the "hearts and minds" programs—the strategy of appealing to emotions and reason to sway a population. Early challenges in counterinsurgency efforts were described as hindered by a lack of understanding of Vietnamese culture and inadequate language skills. Indeed, the foundation of the Civil Operations and Revolutionary Development Support program, jointly administered by the South Vietnamese government and the U.S. Military Assistance Command, was driven by the new focus on a hearts and minds approach as well as an urgency around supplying cultural and ethnographic intelligence to the troops.[28]

In recent years, these trends took on new iterations: after 9-11 and amidst the 2003-2011 Iraq War, the U.S. military—in various ways, across different branches—framed the Counterinsurgency doctrine in part as a softer approach, lauding the beginning of the "Cultural Turn." The Cultural Turn might be read as part of military humanism—the entwinement of humanitarian interventionism and liberal imperialism (Chomsky 2008; DiPrizio 2002), which, although "dressed up in the cloak of humanitarian morality" (Fassin and Pandolfi 2010, 22), is no less war. Senior counterinsurgency advisor to General David Petraeus, David Kilcullen, described counterinsurgency as "armed social work," (Gregory 2008, 13) and soldiers were charged with understanding the "human terrain," "the social, ethnographic, cultural, economic, and political elements among whom a force is fighting" (Kipp, Grau, Prinslow, Smith 2006, 9; see also McFate 2005, for the earliest development of the concept) precipitating the widespread development of cultural trainings for deploying soldiers and the hiring of anthropologists to embed on Human Terrain Teams to supply soldiers with cultural knowledge in the warzone. In an echo of practices long used by the Green Berets, conventional soldiers were trained in cultural knowledge and in the making of rapport—albeit in a far more programmatic and less agile manner than their predecessors. At its height under the command of General David Petraeus, the Counterinsurgency doctrine was criticized and largely fell out of fashion. In 2015, *Foreign Affairs*, a journal that had previously published articles supporting the Counterinsurgency doctrine in Iraq, noted that the strategy was doomed: "The blame lies not with poor implementation but with the strategy itself" (Jeffrey 2015).[29]

Nonetheless, tendencies from this period persist in other forms. More recently, President Obama regularly emphasized lightening America's footprint through the use of drone warfare and proxy armies[30]—progressions that point to something like James Der Derian's ethically ambiguous "virtuous war," an antiseptic mode where killing is separated from dying (2009).[31] While not explicitly espousing the same logic and making for a far louder and bombastic presence, President Trump, meanwhile, radically expanded the use of both drone warfare and Special Operations Forces. These trends all point towards the recent attempts to reframe war's violence (the so-called bang) and make it less visible—or indeed, rather, *less felt* within America's shores—in part by outsourcing it to non-American bodies, both machine and human.[32] These attempts to stage a diminished global imprint, of course, coexist uneasily alongside rising American participation in wars and conflict around the world, including the war in Afghanistan (2001–2021), the Iraq surge (2007), the Afghan surge

(2009), Libya (2011), and more recently, reinvigorated involvement in Iraq and Syria (2014–), as well as escalated U.S. military counterterrorism involvement (and, in particular, drone strikes) in a range of nonbattlefield settings (2015–), such as Yemen, Pakistan, and Somalia. After Donald Trump came into power, these conflicts escalated further and the United States itself became more militarized towards its own citizens. Under President Biden, the shape of this arc remains to be seen. Amidst withdrawal from Afghanistan and the country's full-scale collapse to the Taliban (2021)—and thus the overall decrease of the military's global footprint—drone strikes both in Afghanistan and around the Middle East persist, as well as a commitment to counterterrorism missions across the region and an ongoing commitment to the war on terror.[33]

Meanwhile, to laud any of these "softer" and/or smaller footprint approaches as a way forward is to conceal their violence. Counterinsurgency's logics are driven by both an essentialized notion of the Other and a weaponized version of culture. Indeed, more darkly, the so-called humanization and "application of liberal precepts" of this mode of warfare have further "legitimated war making as a political intervention" (Khalili 2012, 3).[34] In their Unconventional Warfare trainings, the soldiers' enactment of Left of Bang is hushed as they covertly slip into Pineland, court potential partners to overthrow Pineland's regime, and work to replace it with a government congenial to U.S. interests. A major who develops training simulations as well as supervises teams on the ground tells me: "the pine and the pin pass weapons here." The "pin," a metonym for the soldier himself, deflects us into the arms of the trees, seeming to erase accountability. That is: it is merely the pin and the pine who pass weapons, not the training soldiers who are using them.

One night in a later phase of the training, I wait with a group of Iraqi role-playing men acting as local guerillas, who will soon be trained in small unit tactics by the American soldiers. We eat *dūlmā* (onions and eggplants stuffed with lime-infused minced lamb and rice, made by their wives to get them through the long training) in the dark. As the soldiers approach, the pine and the pin pass weapons in a hush; the land just barely creaks, under their boots. They have been preparing to forge relationships, to showcase their cultural attunement. The soldiers are told by the military leadership that when they meet with a guerilla or insurgent they are seeking to cultivate, they should learn to speak as they speak, watch how they hold their jackets and their hands, and as close as they can, be them—honing "techniques of the body" (Mauss 1973) to set their interlocutors at ease and gain trust. A military psychologist who helps them fine-tune these capacities coaches them after the exercise: "Identify the norm of who you're talking to and try to echo it. Little tiny things like that: try to blend in with the guys you're talking to"—that is, render the disturbances of your entry, if possible, less perceptible.

·   ·   ·   ·   ·

Amidst a proliferation of macroscopic, geopolitical approaches to Empire in the years after the 2003 American military invasion of Iraq (Wallerstein 2005; Harvey 2003; W. A. Williams 2006; Zinn, Konopacki, and Buhle 2008; McCoy 2009), there have been markedly fewer more granular, microscopic examinations of the topic. Catherine Lutz called for the tools of ethnography to zoom in on and critique the multiplying sites: "the vicissitudes on the ground" (2006, 594),[35] emphasizing the need for ethnographic excavations of imperial coordinates and logics around the globe, including: "U.S. military bases, soldiers on exercises with and training other militaries, and sex-industry and tourist sites frequented by soldiers. It would include U.S.-AID functionaries and operations; cultural diplomacy as enacted through U.S. embassies . . ." (598). The American military's production and negotiation of difference in its predeployment training exercises offers one such crucial site on the ground. More recently, ethnographic inquiry has begun to fill this void (McGranahan 2010; Simpson 2014; Vine 2015; Dewachi 2017; McGranahan and Collins 2018; Al-Mohammad 2019).

Much recent ethnographic work on the U.S. military and militarism takes the lifeworld of the American soldier as an epistemological entry point. Among this work, essential embodied and phenomenological ethnographic perspectives have turned inward, illuminating the formation and affective entailments of the American soldier amidst the nation's warmaking projects (Macleish 2015; Wool 2015; Wool and Messinger 2012; Finley 2011; Messinger 2010; Kilshaw 2009). *Pinelandia* conversely posits that a study of American militarism is inextricable from its engagements with its others. While similarly seeking to conjure the embodied lives of those enmeshed in military worlds, I instead offer an ethnographic perspective that foregrounds Empire as one of the organizing principles of American militarism. Empire acts as a relation between self and other, constituting the gap (real and/or imagined) between the two. Both a military-political-social-economic set of practices as well as a set of ideologies of domination (which the United States is presently structuring), Empire intrinsically relies on *other* human beings, which it constitutes as both useful and dangerous. Excavating the space in-between, the space of the encounter, this book posits that ultimately to understand and represent Empire (and the wars produced thereby and therein), we must work to understand how the other is being produced (and destroyed) within its warp—as well as the complex lived experience of that other. How, then, does Empire look at itself, and how do others look back, and what occurs in that interface?

Empire here is a "way of life" (W. A. Williams 2006)—embedded in institutions and more intimate everyday practices and part of an interconnected global structure—such that different imperial sites must be read as connected rather than "disaggregated" and camouflaged (Stoler 2018, see also Saleh 2021).[36] Empire thus brings together the entwinements "between settler colonialism, racism, economic hegemony and political interventions" (Saleh 2021, 19),[37] ravaging lives in its wake, creating landscapes as fragmented as project of Empire itself, and producing the very "ungovernability" (Dewachi 2017) that it disavows. In my study, we zoom in on such a nexus of forces, from war and imperialism, racism and orientalism, and knowledge production, and how they meet—in some sense turning the whole world "into a target" (Chow 2010).[38]

Recent studies of America, particular in the field of Native American studies, locate the beginning of American Empire to the origin of the country itself, with its appropriation of indigenous land (Byrd 2011; Simpson 2014). Arguably, the United States, from its earliest origins of settler-colonialism through a long history of slavery and imperial ventures, is constituted via its troubled relationship to otherness.

This book examines the militarization of the human—the turning of the human body or mind into a tool of the military, one of the oldest tactics of wartime—amid increasingly posthuman times.[39] In such times, meanwhile, "even human connections . . . are easily infiltrated by what President Dwight D. Eisenhower once called the 'military-industrial complex'" (González 2016, 15). In the midst of increased development of wartime machine technologies of surveillance and attempted omniscience, the American military has also militarized human beings to produce soldiers as insiders within the warzone. In this logic, a "local" individual assists an American soldier in deciphering adversaries and allies, as well as aiding soldiers in a multiplicity of ways in extending the military's agency on the ground. To this end, soldiers rehearse making relationships and enacting what the military calls "rapport" or "connection" with wartime others, in order to accomplish a range of operational ends—in a complex entwinement between sentiment and power.[40] Indeed, soldiers, too, might certainly be considered technologies[41]—as their bodies and minds are likewise turned into tools of the military; however this book focuses rather on militarized others.

While both military personnel and doctrine frequently referred to "local" individuals as potentially helpful *tools* in assisting American soldiers, when I was in the training camps with role-players, I heard military language that was more explicitly mechanistic. Once, I overheard a major in charge of the exercises describe the cultural role-players as "the Apparatus" to one of his colleagues—a live appendage to the military-industrial apparatus. I stumbled upon the concept that this book seeks to undo and eventually implode. I had already noticed how these people were maneuvered within particular parameters—what the military called "Left and Right limits," as if there were spatial coordinates they were not permitted to transgress. These logics extend broadly in military discourses: the role-players are simply one entity within a bigger apparatus of Left and Right limits. Further, military personnel regularly described these individuals as an operationally useful training tool, a tool requiring limits and coaching to enable maximum efficacy. I offer the analytical conceit of the "human technology" cautiously, offering caveats about the risks of the representation therein. The notion of mobilizing a person as a machine is both dehumanizing and objectifying, a dark apotheosis of late capitalism. However, I believe that the military logics and discourses on the ground bear out this conceptual entry point.

Using Empire's language is a way of asking how Empire looks at and produces itself—as a means of working to resist and undo those logics rather than reinstating them. This book offers also the counter of phenomenology of the lived experience that those structures and logics seek to assimilate. Human beings are not machines, even amidst attempts to extract maximum and prescribed efficiencies out of them, to turn them into repositories of military value.[42] Turning a human being into a technology for a singular use and foreclosing other modes of being and becoming collapses on itself. I thus enter the conceit strategically, thinking first parallel and then *against* a military logic—a logic that is cultivated in policy and national security circles on high (as I show in chapter 1) and implemented in varying ways and degrees by military operators on the ground (as the rest of the book bears out). I critique the ways wartime intermediaries are militarized as tools, while refusing complicity in that reifying discourse in my own representations. Rather, I conceptualize these role-players as ethical actors, who, like others, are alternately confined by structures and making choices in the world. Indeed, if, from a military perspective, human beings can be mobilized as technologies, I counter that these individuals in no way see themselves that way, and they must not be seen this way. Conversely, they are individuals making choices, enacting agency, and getting by within conditions not of their choosing. Employed by the U.S. military as, in some instances, exemplars of their cultures but often ejected to the peripheries as traitors by their own countrymen and as potential spies by U.S. soldiers, these wartime intermediaries negotiate complex injuries, claims for recognition, and varied subject positions.

The Iraqi role-players are both "free agents and embodied conduits" (Inghilleri 2010) for military institutions, interpellated by the military's projects yet also seeking to help Iraq on their own terms (Campbell 2016). Alternately seen as mediators and collaborators, hired as cultural technologies to decipher or embody their cultures but bearing tremendous costs, this work thus acts as a recuperation of the complex subjectivities of a particular population of wartime intermediaries. Amidst the American occupation of Iraq, and the military's Cultural Turn in particular, certain Iraqis were "humanized" instrumentally (and discarded when necessary) while a vast many others were turned into casualty statistics in the American media. This work insists on the complex, rich subjectivities and humanity of these Iraqis, drawing inspiration from crucial ethnographies of wartime Iraq and its aftermath (i.e., Saleh 2021; Ali 2018; Dewachi

2017; Al-Ali 2007) while simultaneously eviscerating a military logic that seeks to turn a person into a tool.[43]

Person as technology (or anti-technology) is hardly a new trope: Aristotle compared animal and machine movements; Descartes sought to sever the two categories to assert the superiority of human reasoning; theorists of political economy culminating in Marx told of the peril of capitalistic assimilation of human beings themselves into the economic apparatus (1867 [2007]; see also Adam Smith [1759] 2002; Taylor 1913);[44] the Frankfurt school cautions that instrumentalizing the person turns him or her into a thing.[45] Using a human as a technology disentangles the body and self from its web of relations, essentially transforming a gift into a commodity and converting human actions into "lines, planes, and bodies" (Adorno and Horkheimer 2002, 67).[46] In his writings on techne, Heidegger warned against man himself becoming a "standing-reserve," a resource like those he sought to tap ([1954] 1977).[47] Foucault reminded us of how bodies are made docile and useful within disciplinary regimes (1977).

Yet if human technology is at least as old as the Greeks in its theorization, its contemporary iteration deserves a closer look: amidst late capitalism in the post 9-11 military-industrial-media-entertainment network (Der Derian 2009), how and at what cost is the person made cultural machine in wartime? And how do these military cultural technologies work or fail—and for whom, and on what terms? Lauded as cultural projects to minimize violence, my work shows a much more complex picture. I interrogate here how cultural knowledges are construed by the American military as mechanisms to further penetrate within the warzone, while querying the limits, contradictions, and implications of that project. I contend that as military personnel seek to become cultural insiders, they are paradoxically often trained in cultural archetypes that ring incongruous for at least some if not many of the cultural role-players themselves. Such incongruency produces moments of excess, such as laughter, for the role-players, who are asked to embody constricted versions of their cultures. These cracks in the simulation offer a reprieve for the role-players, while demonstrating a complex tension for military personnel: as their fixed categories are exceeded, the trainings often, but not always, continue undisturbed. However, as I personally experienced as a participant, when training soldiers cannot "place" events or persons within their schemes, such moments can lead to violence. Ultimately, I argue that the training mechanism not only partially fails on its own terms but also, perhaps most hauntingly, when it does *work*, it offers a window into the production of wartime violence itself. This machinery and its failures allow for a glimpse into military sense-making itself—and its costs.

·   ·   ·   ·   ·

Amidst burgeoning machine technologies of surveillance and targeting in the twenty-first century—and technophilic theorizing among academics and military scholars that war itself was becoming "post-human,"[48] the Counterinsurgency doctrine and the accompanying Cultural Turn were lauded as part of a softer approach by both military and media (e.g., Kilcullen 2006; Packer 2006; Sheridan 2008). Meanwhile, excavating military understandings of "culture" reveal darker logics.

If American military personnel see local intermediaries as potential technologies of culture, "culture" itself is understood variably, acting something like a black box that can be filled with content as needed. In her article, "The Military Utility of Understanding Adversary Culture," Montgomery McFate proclaimed: "Understanding one's enemy requires more than a satellite photo of an arms dump. Rather, it requires an understanding of their interests, habits, intentions, beliefs, social organizations, and political symbols—in other words, their culture" (2005, 43). Culture has frequently been represented in military writings as a mappable entity in connection with the notion of the "human terrain": that is, "those cultural aspects of the military environment that, due to their static nature, can be visually represented on a geographic map" (Holmes-Eber and Salmoni 2008, 14). This notion seems to privilege a more classically structural notion of culture: for example, focusing on the kinship diagrams of a village's patrilineal lineages. Still, in a conversation, one cultural instructor to the military countered: "But even though visual diagrams appeal to the military, you cannot turn culture into a map." She described herself as doing quality control on what the students learned, describing how in the early days of the Culture Turn, "culture was taught through Wikipedia."

In contrast to this structural conception of culture was another military vision, which represented culture as more interior or psychological. At a

Department of Defense summit in 2007, a white paper on the topic, "Regional and Cultural Capabilities: The Way Ahead, Regional and Cultural Expertise," quoted former representative Steve Israel (NY-3), who described culture as something like a psychic force, the motivations and intentions of the adversary:

> Most Members of Congress view Pentagon transformation as a weapons issue, but it is more than that. We need a cognitive transformation. We give our forces exquisite situational awareness using satellite imagery of targets, digital images of friendly forces, live feeds from UAVs [Unmanned Aerial Vehicles] of the battlefield, but what they need more of is cultural awareness. Indeed, they must know the firepower of the enemy, but they must also know their willpower and intent. (DOD 2007a)[49]

In military intelligence contexts, cultural data, mined for discovering adversaries' intentions, appears to be used in particular for its predictive value. A joint intelligence report describes the distillation of various forms of information for prediction: "There will also be a large volume of information concerning weather, terrain, cultural influences, and other aspects of the OE [Operational Environment]. This mass of information can be distilled into intelligence to support a predictive estimate of the situation, as well as adversary capabilities and intentions" (Joint Chiefs of Staff 2013, 1-1).

On the ground at the four military sites where I observed field and classroom trainings, culture was referred to changeably, sometimes focusing on "culture" itself (variably referring to the patterns, structures, intentions of the other) and at other times on "cultural sensitivity" ("do's and don'ts"). Military personnel in the Special Forces community emphasized that they focused more on training cultural "adaptability" and broader models of culture, rather than on a single culture, like the conventional forces. One soldier explained: "We [the Special Operations Forces community] train *models* of culture rather than one specific culture or set of values. We use it as a tool or a model. [The conventional forces], bring in screaming chickens and tell you not to show the bottom of your shoe."

In another military trend in conceptualizing culture, some of the cultural dilemmas within the training exercises posited culture as belonging more explicitly to the private, more intimate sphere. These exercises regularly staged a struggle between cultural relativism and personal ethics (so-called American values), which are generally understood as ideal moral absolutes. For example, soldiers were frequently required to answer dilemmas about how locals treated their women in wartime. They were generally coached to "leave culture alone" if possible.[50] When I asked soldiers why the so-called public sphere didn't also comprise a domain of culture that should be protected from interference, we hit a block. The possibility that nationals might want to self-govern or might not want to cooperate with Americans was generally seen as exterior to so-called culture. In this way, the question of power was a contradictory entity in military understandings: it was both intricately part of culture when convenient (i.e., in order to make diagrams of power brokers) and cleaved from it when it was not.

Amidst these variable understandings of culture, there is one prominent unifying factor. In each case, culture is treated like information. In "The Storyteller," Walter Benjamin describes the form of information as "lay[ing] claim to prompt verifiably." Quoting Villemessant, Benjamin elaborates: "The main requirement is that [information] appear 'understandable in itself'" ([1968] 2007, 89). So too, the U.S. military might be said to understand the tool of "culture" as a form of information: that which is and should be immediately digestible and useable. A cultural handbook used by the Marines and edited by anthropologists referred to "operational culture" rather than to "culture," explaining the distinction thus: "In anthropological terms, culture is the sum total of ways of living built up by a group of human beings that is transmitted from one generation to another. Militarily, however, this definition is insufficient and does not really assist the . . . warfighter because it does not describe reality as it will exist on an operational deployment (Holmes-Eber, Scanlon, and Hamlen 2009, 137).

The manual described operational culture as "a shared world view and social structure of a group of people who influence a person's and a group's actions and choices," emphasizing *observable* phenomena "that can be analyzed and incorporated into operational planning." In this rubric, there were thus parts of "culture" that would not be relevant: "The warfighter is not going to be concerned with all aspects of culture, but only those aspects that influence the area where the warfighter operates" (Holmes-Eber, Scanlon, and Hamlen 2009, 137). This ethos, the need to compress "culture" into its observable operational possibilities, to be used in fighting war, was summed up by a training soldier I spoke to: "You can spend hours talking with them about culture and belief systems. And in the end you get what you need." Via understanding local mores and forging local connections and affective ties on the ground, war-fighters are likelier to secure what they *need*. Put more bluntly by a military instructor to his soldier-students in a class I observed, and reminding us that *both* American soldiers and their Middle Eastern interlocutors are conscripted into the larger apparatus of Empire: "You're learning about culture and cultural understanding in order to be a machine, a surgical machine implementing U.S. policies downrange." He elaborated: "We're not trying to make you into lovey-dovey singers of kumbaya. Hopefully, [this class] will give you the ability to be that much more lethal."

The essentializing culturalist reading of conflict is in no way new. There is a long colonial history of military and colonial rule positioning the "culture" of the adversary/colonized as the problem and thereby deflecting from geopolitics and histories of power (in the Middle Eastern context, see, e.g., Ahmed 1992; Abu Lughod 2002; Ali 2018).[51] For example, the U.S. imperial script in Iraq has generated "a timeless and culturalist interpretation of violence" (Saleh 2021, 31). Reading and thereby producing culture as archaic, oppressive, and static has served as both justification for intervention as well as a means to reduce culture to bullet-points of useful information in wartime—and a means to perpetuate war. Rather than putting an end to war, such knowledge might be read instead as "warfare's accomplice" (Chow 2010, 16).[52]

This study illuminates specifically the height of a single institutional moment in the military-industrial-academic complex, with the 2003–2011 Iraq War as its backdrop. While the story told herein (based on ethnographic research namely between 2011 and 2013, with follow-up fieldtrips through 2015) coalesces around the timing of the Cultural Turn and its entailments, it also invites a broader look into the contemporary practices of American Empire. This includes those entangled within it, while working through the particular challenges of writing about military worlds—and thus "studying up" (Nader 1972).

I make a range of very particular analytic, ethical, and representational choices in this book. First, as a means of protecting the identities of Iraqi role-players, all of whom are potentially imperiled in Iraq for choosing to work with the American military, I took particular care. Although I frequently collected life-histories of Iraq role-players, I do not include them here. My interlocutors were anxious about their potential identifiability. Thus, rather than rendering the thickness of individual lives, I show a range of role-players in context: interfacing with military labor and its complexities. In addition, I obscure or alter particularities of individuals as a necessary measure, in some cases, for example, making composites of multiple individuals. I am indebted to these people, who were caught in a web of war and Empire. They were asked to make difficult choices and chose to work for American soldiers and companies for a complex range of economic and ideological motivations. I am also indebted to many other Iraqis who did not make these choices of affiliation—many of whom critiqued U.S. occupation from the outset—and who helped me make sense of this fraught period.

Second, I call this book *Pinelandia,* as I am describing a "broader archipelago" (Graham 2007) of mock villages and classroom trainings across the United States, rather than focusing on any one alone, and without specifying particular units or exercises. This representational choice is ethical (designed to protect the identities of Iraqi role-players and American soldiers) as well as analytic. Most of my fieldwork took place from 2011 through 2013, a time when "culture" was a buzzword across many military training settings. I witnessed a wide range of trainings and subcultures—including Special Operations Forces, conventional forces, and a military training academy—with considerable differences in complexity, scope, and stakes (e.g., in the military academy, the cadets were not immediately deploying). The key exercises I focused on were, however, similarly "cultural," with an emphasis on making the "human terrain" legible.[53] In each instance, soldiers were taught to differentiate between and fix social categories in the warzone as well as secure allies and defeat adversaries, and to take the glaring edge off their presence through demonstrating to the locals some degree of cultural savvy, pathos, and interest in their circumstances. The Special Operations Forces and conventional forces have essentially different doctrinal differences and missions, and those with military ties may thus dispute these representational choices. However, I contend that there was enough overlap in the kinds of exchanges that occurred between trainees and role-players to merit the analytical choice of examining these exercises in tandem. Certainly, this cultural moment impacted military personnel differently, from junior enlisted soldiers to small unit leadership to commander; indeed, not all personnel had responsibilities to liaise with the local population directly. Nonetheless, the Cultural Turn impacted, at least diffusely, everyone deployed during the Iraq War. That is, a shared logic from this institutional military moment ultimately binds together the exercises and participants, however dissimilarly.

In this process, I refer to a cluster of different programs and initiatives that are not in any way synonymous, but they are part of this moment. For example, the Minerva Research Initiative is the U.S. Department of Defense's 2008-launched program, which supports, through unclassified grants, university-based research in the social sciences. As specified on its website, this research is "aimed at improving our basic understanding of security, broadly defined." The Human Terrain System, fully launched in 2007 and shut down in 2014, was a U.S. Army Training and Doctrine Command program that employed social scientists to provide soldiers with an understanding of the local cultures they were deployed among.

Further, while the "Pineland" of the Special Forces receives focus at various points in the book when I specifically refer to the Green Berets (especially in the introduction and chapter 6), most of the text focuses more widely on a range of patterns I saw across military bases, among different units, across both Special Operations Forces, the General Purpose Forces, and military training academies.

It is important to note that the overarching training for conventional units and special units is certainly widely variable. For example, during the height of the 2003 Iraq War, soldiers in the conventional Army, after their ten-week basic combat training and variable advanced individual training depending on their military occupational specialty, would then receive cultural and tactical field training for fourteen days or less alongside basic language training for their zone of deployment. In contrast, at the time of writing, the Green Berets' full qualification course lasts between fifty-six and ninety-five weeks, and typically includes four to six months of language training and a final field training of four weeks, which includes more complex and contingent exercises (however, the order of the language training changed several times over the course of this research). Moreover, linguistic (Arabic) capacities among the training soldiers were somewhat variable; however, on the whole, the soldiers used an interpreter during the exercises, creating some degree of uniformity.[54] Regarding linguistic capacities, I did not encounter any instances in my observation where a training soldier spoke strong enough Iraqi Arabic dialect to catch the nuances of what was said either during a simulation or while chatting with role-players. I never saw soldiers trying to answer role-players in Arabic or reply in English to something a role-player said in Arabic. I witnessed one more advanced exercise where soldiers were required to practice Modern Standard Arabic, and I specify this in the text. Thus, on the whole, Arabic produced something closer to a cultural or "atmospheric" backdrop—and in some cases, a source of bewildering "noise," adjacent to other forms of noise (the call to prayer, ambient city noise, explosions).

Without desiring to *flatten* the vastness and particularity of military subcultures and exercises, I contend that there is a shared trend worth examining: namely, the use of relationships with the "local" population as a means of minimizing U.S. presence (namely, harm) on the ground, while accomplishing operational goals. Ultimately, amidst considerable differences across the U.S. military, key resonances between the cultural trainings chimed, and the representational choice is both analytic, a means of shining a light on a broader cultural moment, and a means of protecting identities.

Finally, for a book that straddles many mock villages in different settings, *Pinelandia* is largely set in the woods—villages that are contained and in some ways constricted by those branches—however, this is in part a convention for narrative unity. The sequence of villages described were often in woods, yes, but also sometimes in fields and in deserts, and always surrounded by military cubicles and interchangeable rooms decorated with American flags.

.   .   .   .   .

One afternoon in the very first months I was conducting fieldwork in 2011, a captain asked me for my opinion of one of the exercises he had just run. We had been chatting for a while—the topics ranged from the time we had each spent in the Middle East, to my impressions of this Army town (it was the first one I'd visited, and I couldn't believe how many boot repair shops there were), to what I had studied as an undergrad (creative writing and French literature)—and the mood was casual and relaxed. Amidst this stream of dialogue, I said to him: "Well, to be honest, I was surprised that it seemed so rigged." When he asked for more clarification, I offered a few particulars. Several days later, I returned to the training site to see that the captain had *implemented* my off-the-cuff replies to his query, thus launching for me a series of uneasy insights. First, the captain had used one of the techniques of the Cultural Turn *on me:* building rapport in order to secure information. And second, I was moving through a dangerous space where the co-optation and weaponization of cultural and regional knowledge were the goal: I myself was seen as yet another potential repository of that knowledge, as an anthropologist of the Middle East. In subsequent interactions with military personnel, I took great care with each reply, refusing to become an unwitting contractor—even while knowing I was already moving within a structural logic far bigger than I was, Empire, where it was impossible to decide whether or not to opt in. This all came to a head over time, as I was given opportunities to play various roles in the simulations. And with my last role, Gypsy—the girlfriend of a slain guerilla—the project essentially came to an uneasy close.

Access into these fraught military spaces came incrementally and was likely born of a particular geopolitical moment: the military was strategically engaged with "culture," the province of my own academic discipline. I located myself as an outside observer who was interested in future academic jobs and did not plan to work for the military. Nonetheless, as I sought to understand the position, discourses, and logics of military personnel, my military hosts were likewise interested in trying to understand those of the visiting anthropologist. My access was contingent on my non-disclosure of any "information that could provide details that persons or parties with interests hostile to the United States may find useful thereby imperiling both military and civilian lives." The ground rules asserted further that they were "in no way intended to prevent the release of embarrassing, negative, or derogatory information" (U.S. Army John F. Kennedy Special Warfare Center and School 2011). I believe I have hewed to these terms: I observed exercises that were unclassified and signed a document that if I were exposed inadvertently to classified information, I would be notified and told I could not disclose the material. On one occasion, a major asked me to exclude a witnessed detail from my writings for security reasons, and I have done so. This was the single time I was asked to not include something that I observed. In addition, these exercises I describe are variations on prototypes that have been analyzed in previous studies.[55] Core themes and preoccupations repeat themselves in the simulations; for example, how to map the human terrain while procuring intelligence; how to gain rapport with the locals; how to enact cultural sensitivity and how to adopt an attitude of cultural relativism, when the locals do not align with military-perceived American values.

The militarization of local intermediaries to embody Middle Eastern culture must also be located more broadly within the American military's interest in the figure of the *subject matter expert,* and in particular, the anthropologist. As I have described, in the military's Cultural Turn, this range of figures should ideally assist the soldier in deciphering the adversary and securing allies; embodying (the role-player) or explaining (the anthropologist) cultural patterns; and generally becoming a knowledgeable insider in the warzone. During my fieldwork, I was continuously reminded of my own uncomfortable border location: an anthropologist in training who insisted on independence and was constantly dodging military personnel's attempts to procure my professional advice on how to improve the simulations. The project became increasingly "para-ethnographic" (Marcus 2000), evoking the violence of ethnography as a knowledge-making practice. I felt most uneasy as I experienced the kindred tools that soldiers and anthropologists use to comprehend and, within limits, to find our feet and become more at ease within an unfamiliar setting. As the soldiers were trained to *wait* to procure operationally relevant information, in order to not estrange their informants, *I, too,* watched myself wait to ask both American military personnel and Iraqi role-players more sensitive questions. In each case, we were steadily trying to accrue the currency of trust. In each case, we inhabited certain locations of power. Although I was seeking to understand a social phenomenon as a scholarly pursuit and not to further war-making, I still felt uneasy about the tensions in the role. It is one of the complexities at the center of this story that the very people who are experts in culture—academy anthropologists—are those with the least control over how it is used. As a foil, this book is likewise populated with military anthropologists—those willing to and paid to instrumentalize cultural knowledge for wartime use.

Meanwhile, I watched the training soldiers practice composing useful dispositions and mirroring their local interlocutors, while taking critical notes about these instrumentalizing logics. I simultaneously composed a professional demeanor, that of an academic observer. I carried a notebook and a digital recorder. Yet, when I interviewed the American military personnel about their experiences in war and about architecting the predeployment simulations, I noticed I began to "shoot the shit" and joke around—about everything from American pop culture to the towns that surround the military bases. They liked having a woman around. "Nomi Stone!" they would call out from the other side of the fake mosque. "Are you going to put me in your book?" I was an academic with a notebook; but in some moments, I watched myself begin to assume the location of American chick hanging out with American soldier dudes.

Having no prior experience with American military personnel, and having spent the last fifteen years either studying remotely or doing research within the Middle East, I was far more comfortable with the Iraqi role-players. But when I interviewed the role-players, didn't I morph as well—perhaps if only into a state of being that afforded me ease? The first day I met the role-players at one of the military bases, the female role-players were in the fake market, hawking *Lablābī*, that chickpea soup sold in paper cones that you can buy in the Middle East. I replied, in Iraqi Arabic: "*Wīn a-lablābī u ashgid*" [Where is the *lablābī*, and how much]? The role-players were so stunned that I spoke Iraqi dialect that, during the interlude between simulations, they sat me down and gave me a Coke. When I told them my name was Nomi and joked to them that it was just like those limes that grow in Basra—like that tart and pungent flavor found in so much of Iraqi cooking—they laughed with recognition and began to call me "Nomi Basra."

It is certainly a bit ironic that a book intended as a critique of the use of "culture" as a medium of domination moves continuously and necessarily through descriptions of culture itself. Most ostensibly, we have "Iraqi culture" as seen by the military with its diagrams. Then we have Iraq as seen by myself, an anthropologist who is not of Iraqi heritage and who has only spent time with Iraqis in diaspora, not in Iraq itself—through the sieve of years of reading Al-Sayyab poems over cigarettes, making *dūlmā*, talking about Iraqi politics, and shopping for wedding dresses with Iraqi friends. Even amidst my sympathies with Iraq, an Iraqi friend reminded me that my unexpected Arabic fluency surely "punctured the seal of [the role-players'] privacy" on a military base. We also have military culture as seen by Iraqis, and military culture as I see it (as a person with no background or family history in the military), with all its insignias and blazers and buttons, with its pine-needle tea, dip and slouching. In his poem, "The Disappearances," Vijay Seshadri asks: "Don't you understand that history is being made?" And then continues: "This is you as seen by them, and them as seen by you, / and you as seen by you, in five dimensions, / in seven, in three again, then two / then reduced to a dimensionless point." Amidst such vanishing points, I try to be aware of the reifying traps of even more carefully executed cultural description and the traps of my own disciplinary medium of analysis. In writing about other ethical actors in the world, I continuously work to both hold myself to that standard and to scrutinize the way in which I likewise move through the hall of mirrors. Indeed, using academic language and quotes is an academic form of translation—and its own simulacra.

Amidst the traps and mirrors and power structures of this fieldwork—as well as of the anthropological method itself—I was most inspired by an "ethics of being-with" (Al-Mohammad 2010), wherein "the cord" between lives "the line, the relationship, the *with* of being [is] ethical itself." Lives are and become enmeshed with others, and this entanglement bears tremendous responsibilities. I have sought to represent those cords and my ties therein with care.

I conducted this research as the culminating academic exercise to become an anthropologist. Yet as I formulate my observations and critiques about the U.S. military's tactic of attempting to become insiders in order to acquire information, I struggle with my own complicity: I dissolve between two languages, taking notes in my notebook. I am observing the workings of the most charged sites of power, within American Empire, as American military presence definitively sprawls across every corner of the globe. I hunt every day, within anthropology, for the tools to critique the co-optation of anthropology for a means to reimagine and redeem anthropology. The anthropology I had been trained in and so valued had been neutered by the military; it collapsed in on itself when devoid of its critiques of power, an analytic mode I found central to contemporary anthropology.

Yet the ethical and political stakes are so much higher than any conversation confined to the discipline of anthropology. According to Theodor Adorno and Max Horkheimer, the ethos inherited from the Enlightenment and intensified in capitalism treats "human actions and desires" like "lines, planes, bodies" (2002, 67). The result is nothing less than "the withering of experience," as categories of existence and being are ossified, refusing the inevitable human remainder—culminating in the greatest ethical peril, of "treating people as things" (Adorno 2005, 40, 42). The U.S. military's Cultural Turn has maneuvered encultured bodies as such, as they act out fighting, bargaining, mourning, and dying on a loop within military parameters, as the "reified, hardened plaster-cast of events take the place of events themselves" (Adorno 2005, 55). Further, these training camps might be understood as one slice of a larger national project that normalizes war as a permanent and indefinite state, one we must question and estrange. This work is a journey into the contours of those violences—and each chapter, cumulatively so. The shadows of the trees grow around us.

[Field Poem Fragment]

They call it the kill zone   I am going deeper in
We haven't slept   Pollen burns my nose

[Field Poem Fragment]

Sunlit and dangerous, this wooded road.
We are follicle and meat and terror and

the machines leave their shells naked
on the ground.

1   The Making of Human Technology

As I drove into the town adjacent to the base, I passed signs for military discounts at the movie theater, the golf club, the Arby's, and the AT&T store. There was a billboard for Black Ops Paintball, which invited church groups and bachelorette parties to *Blast your friend with FUN*. Past barbershops, boot repair shops, the Applebee's, and beige miles of on-post military family housing, I met a soldier named Jason in a chain coffee shop with large plush chairs in a strip mall, and we both ordered lattes. A captain who had recently acquired his Green Beret and finished Robin Sage, the final training exercise in Unconventional Warfare, Jason described to me the enormity of the challenge of seeing, understanding, and acting while covertly deployed in a denied territory. He explained: "How does someone accomplish something from a hideout? He's not supposed to be there. He doesn't have the strength to do things himself. It's like being trapped inside a box. You're blind. You can't see anything. You can't hear anything."

Yet, Jason explained, this box did not incapacitate the American soldier: "You're in a box, yes, but people can come in and tell you things. You can't go outside by yourselves; but you've got to *make stuff happen!*" Indeed, "stuff happens" via the wartime intermediary. Jason emphasized the need to train his local counterparts—what soldiers call "indig"—in methods that they would have access to even in the complete absence of U.S. forces: "Scale that out to a stay in a country. You are blind. Once the U.S. withdraws, you have ISR [Intelligence, surveillance, target acquisition, reconnaissance], but if we don't give it to our indig—we need to train them in methods to counter the enemy that don't involve technology." It was possible to leave the imprint of American methods and operational goals—to *stay* in a country, from afar.

Jason is one operator, a single soldier, interpreting and working to enact a military architecture that was produced by policy circles and power brokers on high.[1] I first met him at a summer picnic at the Special Forces Association (SFA) several weeks prior, celebrating a new class that had recently completed Robin Sage, the Green Berets' famously difficult culminating exercises. In Robin Sage, the students must practice Unconventional Warfare with role-players who take the parts of guerilla chiefs and the indigenous guerillas whom those chiefs must train, during a ten-day stint in the woods. The guerillas are ultimately trained to fight on behalf of American aims, overthrowing their own country's government and instating one more congenial to U.S. interests.

On the day of the picnic, the newly minted Green Berets ate barbecue in the SFA's well-tended lakeside bar with dozens of earlier generation Special Forces guys, many of them now retired. We were miles away from the grimy stretch of road with all the fast-food chains and barbershops. Out the window, an American flag rose over the windy blue day, a small bridge, decorative fountain, a lush row of trees, and a pond. "That pond is full of cremated SF guys," a soldier later told me. Cars, some with U.S. Army Special Forces' license plates, adorned with a beret and a knife, were parked in rows. Wives, hair done, lipsticked and smiling, spooned baked beans and potato salad onto their children's plates as sun sparkled through the windows. Older Green Berets teasingly asked if the new soldiers had enjoyed their "pine needle tea," a staple during the training in the woods— boiled directly from the scatterings on the forest floor. The new trainees were relaxed and confident, responding cheerfully to the older cohort's provocations. The new guys had at last endured the trials of Pineland and now could join the Special Forces club as real Green Berets about to deploy. There was an intimacy in the room, and an excitement and feeling of readiness on the part of the newly trained soldiers to actualize what they had learned. Beyond, out the window by the lake, were acres and acres of pine trees, and beyond them, a globe boiling with lands at conflict, lands these young men might one day rearrange.

Later that week in the coffee shop, as we spoke of what kind of seeing, knowing, and understanding might be possible in wartime, Jason returned us to the war game. He explained: "Pineland is that blind box. If you're given a complex problem that your [Special Forces] team can't perform . . . if you get caught outside . . . it's game over! So you have to act through surrogates to do something you can't fully comprehend." Jason explained that in these challenges: "You're trying to get a grasp of the operation and strategy and the country through someone else's eyes. You can't understand that country or culture or language, even with years of study. How do you develop mechanisms in this position?" The blindness Jason describes is multiple: a physical blindness from being in denied territory wherein the soldier literally cannot discern where and how things are happening, and a cultural blindness from never being truly an insider within another culture. Wartime understanding is specious and fraught with constant doubt. Yet, he explains, with the assistance of a local on the ground ("human technology") Jason is less afraid and no longer sees himself as impotent and enemy territory as inscrutable.

In the war zone, the Other—foreign enemies and citizen populations— have often been overtly dehumanized and subjugated. Here, a more complex dynamic evolves: the Other is cajoled, asked to be a partner, but more specifically, a tool for seeing, knowing, and acting. By using the Other as a tool this way, the soldier dreams of thereby partially disappearing: to be robustly within and among the flows of the world while leaving a more minimal trace. Gilles Deleuze and Félix Guattari muse that if human beings have a destiny, it might be to "escape the face, to dismantle the face and facializations, to become imperceptible" (1987, 171). This longing, in the theorists' case, is the dissolution of the self—the ego, the language-bearing subject—into the cosmos of plants and animals and minerals, culminating in at last becoming *imperceptible*. In this dream, the different forms in the world shape-change, opening new spaces for contingency and possibility and releasing us from the yoke of a fixed and single identity. Yet to try to become imperceptible among other human beings in wartime—

where identity and the grounds of political, ethnic, and national difference are the basis for killing or letting another person live—conversely may contort into a dystopia, a political nightmare devoid of ethics. Who is becoming imperceptible to whom, we might ask, and to what end, and to what ultimate cost?

What the U.S. military describes as "softer" approaches to war—war outsourced to indigenous forces, the use of local intermediaries to help U.S. soldiers mute the edge of their presence, and the Cultural Turn as a whole—are all part of a project to render American forces *less* perceptible in the warzone. However, there is a see-saw effect here: as a result, the locals who work in their service often become *more* perceptible and thus more expendable. Human technology tells us about instrumentally summoning common humanity in wartime as a war tool; this tactic might ultimately be better understood as a masked form of violence.

Indeed, in writing about war and social suffering, theorists have often turned their attention and cautions to processes of dehumanization and dis-identification (Scheper-Hughes 1993; Butler 2004). Namely, the work of war is seen as one of "de-realization," rendering others who are already unimaginable further faceless, and therefore expendable and ungrievable (Butler 2004). Post 9-11 military trainings offer a haunting foil to this narrative: the Other is made more "real," yet namely for a particular American operational end. The Other is made, at once, more visible and more susceptible to harm—thus in some sense, "humanized" as a means to an end—an inhumane end.

·   ·   ·   ·   ·

Moving between an archipelago of military outposts around the country, the sprawl of beige and gray, the American flags, groomed lawns, and insignias begin to become familiar and interchangeable. On the periphery of the bases, it is generally the same as well: the ragged businesses and bars with their pool tables; on the inside, the tidy cubicles. We move here from the operators—the soldiers on the ground like Jason—to the domain of the theorists and architects: the thinkers of counterinsurgency, of the Cultural Turn, and ultimately of the policies that produce human technology.

I take the train to a brick-built military base where a center for operationalizing culture is housed in a portable annex. It is humid. I walk past the office of a Human Terrain Team anthropologist who was embedded with and offering the military expertise, the walls crimson with Afghan weavings and hung with military medals. I pass rooms full of whiteboards and desks for the many different regions of the world where the American military has a presence. In one of the regional alcoves, decorated brightly with local memorabilia (rugs, tapestries, beads) from countries at war, is taped Psalm 91:9–12: "If you say, 'The LORD is my refuge,' and you make the Most High your dwelling, no harm will overtake you, no disaster will come near your tent." I then meet "Natalie," an anthropologist who runs culture classes for the training soldiers.

> "So," she says self-referentially, "you're getting to meet the traitors. The defectors," she says. "How about that?"

> "I'd like to hear," I stumble—then offer carefully—"about your experiences, about the choices you've made."

> "You've certainly chosen a dangerous topic," she laughs.

After she takes me on a tour of the center, she introduces me a gregarious colonel who presides as its head. When I tell the colonel I want to know more about how the military conceptualizes culture, he replies: "Culture is an iceberg. The tip—the part you can see—is the do's and don'ts, manners and courtesies. But if you go deeper, it's how they think, how they make decisions, their reactions. You can only edge beneath that tip. You will never truly understand how they think."

Military personnel frequently use this iceberg metaphor when trying to explain cultural understanding, and even its potential limit points. I am uneasy as again culture is turned into a concrete object, a thing rather than a set of processes. Yet before I can reply, Natalie whisks me off to lunch. We head to a little restaurant that smells of tuna melts and grilled cheese, and nearly all those eating are in uniform. There is a French-speaking cashier, and Natalie speaks in French to her. We sit down to our sandwiches, and I ask her what it's like to teach the soldiers culture.

"To teach the soldiers, you have to get how they absorb information, how they think. But even though visual diagrams appeal to the military, you cannot turn culture into maps," she explains. "You have to find something in between." She explains that she has been something like quality control on what the soldiers learn about culture. That things have changed a lot.

"What was it like before?"

"In an earlier day [in the beginning of the 2003–2011 Iraq War] culture was sometimes taught through Wikipedia."

In 2002, the eventual writer of the famed Iraq counterinsurgency manual, Lieutenant-Colonel John Nagl, historicized the coming turn to culture, lamenting the ultimate inability of the U.S. military in the Vietnam War to locate insurgent foes: "the U.S. Army could not conceive of a war in which its weapons and organization could not only not destroy the enemy, but usually could not even find or identify him" (2002, 198). In 2005, Montgomery McFate and Andrea Jackson coauthored a paper proposing that anthropologists embed within military units to help soldiers fill gaps in comprehension of local population and culture in Iraq and Afghanistan (2005). The following year, the Army and the Marines coauthored *Field Manual 3-24* under David Petraeus, reconceptualizing the 2003–2011 Iraq War as a counterinsurgency (COIN) and laying the seeds for "culture" as the new hermeneutical mechanism for separating insurgents from the general population. The COIN doctrine focused on the necessity of soldiers' "boots on the ground," in order to decipher social actors, separate insurgents from the general population, and woo the uncommitted masses known as the "fence-sitters" to the side of the Americans.

As part of this effort, David Kilcullen, senior counterinsurgency advisor to General Petraeus in Iraq, coined the term "conflict ethnography" (2007), which he proposed would enable the U.S. military to understand the connection between Iraqi insurgents and the broader Iraqi population. "The more unconventional the adversary and the further from Western cultural norms," McFate explained, "the more we need to understand their society and underlying cultural dynamics" (2005, 48). These "underlying cultural dynamics" were converted by the military into manageable cultural categories that could be memorized and applied by soldiers.[2] "Tribe,"[3] "Islam," and "Gender," and in the Iraqi case, "Sect"—what elsewhere Lila Abu Lughod has called classical anthropological "gate-keeping concepts" from the colonial period through the present—were adopted, simplified, and rendered operationally relevant to the Iraqi effort (1989)[4] effacing their historical particularity.[5] These categories were produced in concert between Neoconservative policymakers, military thinkers of culture such as McFate and Kilcullen, orientalist writings on Iraq by academics that reproduced prior British colonial analyses, and also via sectarian discourses circulating among Iraqi opposition groups, who had fostered collaborations with Neoconservatives and the Bush administration in the United States.[6] The reification of these cultural categories and the misappropriation of anthropological tools would become the hallmark of the Cultural Turn.

Natalie had a bit sardonically, a bit ruefully, referred to herself as a traitor, a defector, to me, for having provided such anthropological tools to the military. And indeed, many scholars have critiqued the use of culture as a weapon system. In 2007, the American Anthropological Association condemned the deployment of anthropologists within Human Terrain Teams in warzones, asserting that it violated the AAA's Code of Ethics—and many anthropologists have since critiqued anthropology's collusion (e.g., Kelly 2010; González 2009; Price 2004; Price 2008; Price 2016).[7] Our conversations were thus infused with a complex charge: her demeanor alternated as I perceived it between a bit of self-mockery, defensiveness, and disappointment, wanting to make me—an anthropologist within the academy—see why she had made the choices she had.

In a basement office, mostly monochrome, but decorated with stars and busts and maps, I asked Natalie about the genealogy of the cultural trainings she participated in, and she told me that [the military personnel] were "building the plane while they were trying to fly it." She added that the sign on the door to her office had changed any number of times: "First it was 'Culture,' then 'Anthropology,' then 'Anthropologist.' [The other anthropologist and I] joked that that we felt like the [military] was feeding and caring for their pet anthropologist."

When I asked Natalie how cultural trainings relate to the programs of counterinsurgency, she replied: "COIN and Culture training are *not* inextricably linked. They are two parallel trajectories that find themselves walking in the same direction." Natalie explained that in fact, the early cultural trainings were a reaction to U.S. military blunders in the early stages of the 2003 Iraq War. They were intended to quell American media and citizen outrage over war crimes such as the U.S. Marine massacre of Iraqi civilians in Haditha in 2005 and the U.S. Army abuses and torture in Abu Ghraib: "When the culture stuff started, it was not in response to COIN. It was in response to the horrors of Haditha and a way to placate the U.S. media. Early predeployment stuff was manners and customs." Despite these disparate origins, Natalie explained that COIN and culture thereafter became conjoined: "But David Kilcullen became the guru of all this, and brought COIN and culture together. Like, we need culture to deal with these weird insurgents." Indeed, the phenomenon soon became sexy: "Different people were grabbing onto it. Starting in around 2005, it became a cash cow. All these different culture babies were sprouting. Everyone was posing as a cultural expert."

Melissa, another military anthropologist on site also sought to extricate culture discursively from counterinsurgency, cautioning that in COIN, cultural categories were used to court the courtable and kill the uncourtable: "COIN, seriously? It's a buzzword to throw money at, it's not a way to understand. [COIN] is dangerous!" Melissa exhaled with frustration: "The problem comes with its notions of predictability. That you can categorize people [good, bad, fence-sitter] and sprinkle pixie dust on them so you make them like you; or, you kill them." She pulled her chair towards me as if sharing a confidence and then explained, "This work is complicated. Things don't fit in tidy boxes."

In another office at another base with its whiteboards, maps, and pictures of the Middle East, Mary, a social scientist who develops military cultural curricula, welcomed me warmly into her office. In describing her trajectory, Mary told me she had studied the arts and was passionate about film, but had eventually segued into international relations in her career. A bit of an outsider, artsy, with a liberal background, and interested in trying to understand the military, she told me that she worked to offer her student-soldiers not only regional preparation and practice at ethnographic methods, but a journey towards "introspection and self-consciousness."

Mary explained: "I talk about their impediments to observation, their stereotypes, their xenophilia, their xenophobia. I help them think like an Afghan. How to understand their counterparts."

I asked her how she created such a space with the student-soldiers, and she replied: "I need to reach my audience. If they need to use the f-word, and their dip, and they slouch in their seats. I say, 'I respect you guys.' I left my job to do this. I shock them, and I let them be themselves."

In the course of our conversation, Mary and I talked about our respective backgrounds in the arts, cracked jokes about military acronyms, and she asked me in depth and with genuine interest and warmth about my interviews with Iraqi interpreters. She complimented my cowboy boots. I noticed that she and I both had long (a bit unruly) hair, amidst these polished, tidy hallways where women nearly categorically wore tight buns and suit jackets. There was an affinity between us, and amidst this feeling, I finally asked her, my voice shaking:

"But isn't all this hearts and minds stuff about getting information about adversaries, about another way to target people?"

"It's a quagmire," she replied. Her eyes looked sorrowful for a moment "Yes, [the hearts and minds] is also about active intel[ligence] gathering."

We couldn't find answers in that hour, but Mary hugged me affectionately when I left: "Keep me posted on what you're up to, girl!"

Amidst these doubts and cautions voiced by the military social scientists I spoke to, the "culture babies" continued to "sprout" in various operational permutations in support of counterinsurgency trainings. In 2007, the Department of Defense (DOD) held a summit to discuss the increasing buzz around "culture" explaining in their white paper: "Our vision is robust national defense strengthened through the application of regional and cultural competencies as integral capabilities of the 21st Century Total Force" (DOD 2007a, 4). Subsequently, the DOD continued to hone its *Defense Language Transformation Roadmap* Initiative, which it had launched in 2005, allocating more than $750 million over five years to increase the number of military personnel with language skills.[8]

As the 2003–2011 Iraq War progressed, the DOD increasingly codified cultural trainings and programs that would encourage "cross-cultural competence."[9] Indeed, it is estimated by the Government Accountability Office (GAO) that the DOD spent approximately $266 million between 2005 and 2011 providing general purposes forces with cultural and regional training ("Defense Language and Culture Training," GAO 2011). In this period, the military developed and publicized new cultural trainings compulsory for all deploying soldiers, within mock Middle Eastern villages at the National Training Center at Fort Irwin, California, and at the Joint Readiness Training Center at Fort Polk, Louisiana. In 2008, the DOD developed the Minerva Initiative, to facilitate cooperation between military and area specialists in universities, particularly those who worked on Islam, Iraq, and China. Meanwhile, military funding ballooned to hire indigenous cultural experts and training programs received funding if buttressed by knowledge of "local" experts from "adversary" cultures. Rather than imagining a singular linear trajectory, we might see this moment as a bigger institutional broth: of new programs, experts, and trainings, as well as an invigorated focus on certain military strategies across branches and units that required cultural knowledge. Certainly, this period was full of variegation and change. For example, it is my ethnographic impression that over time, the more simple, static rubrics of adversaries in cultural trainings began to incorporate more instability and unpredictability—rather insidiously, in some cases, coming to mimic the British colonial trope of the "wily Indian"—an ascription of culturally essentialized irrationality and unpredictability to the colonial subject seeking to "outfox" his ruler (Blackburn 2006, 143). Indeed, in 2014, one internal critique, Kerry Fosher, an anthropologist within the defense/security community, posited: "The Department of Defense (DOD) and intelligence community (IC) may prefer solutions that presume static, unchanging culture groups. Yet people around the world do not hold still for the convenience of cultural knowledge databases, field guides, and just-in-time training" (Fosher 2014, xvii). However, even when some military programs came to presume that adversaries were not "static," a conviction seemed to remain that they were ultimately knowable and containable. Within this institutional broth, human technology was cultivated, at terrible lived cost.

·   ·   ·   ·   ·

One bright and cloudless afternoon in 2013, in a beige room embossed with maps on a military base, "Lieutenant-Colonel Williams," an avuncular and precise man who had written monographs theorizing counterinsurgency in the early stages of the 2003–2011 Iraq War, described to me the work of seeing in wartime. He had been deployed in Iraq more than once, and he was invested in the strategies that had been undertaken there. He narrated the visual stilling of a city street in Iraq as something akin to watching a play: "As the war went on, we started to see new things. We were on a street in a village one day and it started to look like a play to me, a stage, and things just did not look right." He explained that as the scene of the war became more complex (referring to increased Iraqi resistance as well as increased sectarian violence, from 2005 through 2006), how important it was to read these visual clues: "There were no children on the street. Two men were carrying something. There were some other guys they were gesturing to. We had to read it."

Soon after, I had breakfast with "James," a theorist of war and war games and an intelligence officer who I had met at a war game conference. He gesticulated a great deal and spoke quickly as he laid out the terms of war preparation. James told me: "Any form of unpredictability can be incorporated if we do it right. The world is only fundamentally unpredictable if you make a category-error in the sense of Gilbert Ryle." James articulates here a brazen military paradigm that I often heard: the world, complex though it is, ultimately can be domesticated. Essentially: you can figure out how *they* think and out-game them. James refers here to none other than Ryle, the philosopher at the center of Clifford Geertz's argument in his canonical essay, "The Interpretation of Culture," taught to many first-year anthropology students. Ryle writes about the "wink" as a meaning-laden moment, but one that might be misread; what if it is a twitch, or a conspiratorial gesture, or a parody of a conspiratorial gesture? For Geertz, an unsituated gesture, which is shorn of context (or "thick description"), is only the compression of the eyelid, and it cannot be accurately read. The gesture, once contextualized, however, may function as a "speck of behavior, a fleck of culture" (1973, 6). Likewise for James, any act or moment—wink, blink, or otherwise—might be comprehended if it is placed within the proper categorical framework. Once an object is assigned to its proper category, one might make "sense" of it and thereby contain it.

Yet amidst these confident assertions, inviting the "reading" and "categorizing" of the world, perhaps there is also a tacit military awareness of a gap between their seeing and their actual knowledge and understanding. As the optical reach of the U.S. military expands within a twenty-first century posthuman technoscape,[10] access into the interiors of adversaries and their cultures remains anxiety-producing for the security community that seeks to "secure life from the species to the population to the individual to the microbe" (Masco 2014, 19).[11] Indeed, the "self/eye, the I of the United States" amasses knowledge about the Other to render that Other a target field at the other end of that gaze (Chow 2010, 16). Mobilizing culture and local individuals as a tool into the Other's intentions and ways of being in the world are yet another militaristic tactic to attempt to magnify that encompassment[12] and balance imperatives of having one's boots on the ground while diminishing American harm.[13]

One afternoon, Lieutenant-Colonel Williams—who had described watching a street in Iraq as akin to watching a play—told me about the potential role of local Iraqi interpreters in deciphering a wartime scene or event. He described his deployment in the early years of the 2003 Iraq War. When there was a moment that was previously inscrutable to him, the interpreter "helped me understand . . . just because the guy has got the robes doesn't mean he's the power-broker . . . that's what I thought when I first came [to Iraq]." With the help of his interpreter, he came to understand who was in charge behind the scenes.

> "Can you tell me more what you mean?" I asked the lieutenant-colonel.
>
> He elaborated: "What I learned to do [instead] was watch the other guys in the room. And it's the guys who are *not* the Shaykhs . . . in this case, the *doctor* . . . who keeps quiet and rarely ever says anything."
>
> "Ah—tell me more how you understood the room?"
>
> He continued: "I realized after a while: *this* is the guy who dealt with the police. He would always turn up for meetings with the police. And I was like, why is the doctor turning up for these leadership meetings? He's the doctor. . . . Well, he's also the Islamic Party's bellybutton."

The lieutenant-colonel explained that he came to understand, with the assistance of his interpreter, that it was those who were most visibly performing who were diverting him away from the operationally relevant information. He went on, explaining that he put all this to the test, watching how various social actors reacted to him:

> "That's how I figured it out. I'm sitting here listening to all the Shaykhs do all their theatrics, and I stood up and I says, *Let me tell ya: I've heard what you've got to say, but this is my rules. And I'm putting my foot down.*"

> "And then?" I asked.

> "And then I watched body language in the room."

Thus guided by his interpreter, the lieutenant-colonel realized that yet another layer of power, which was initially invisible to him, coalesced within the doctor.

> He explained: "The doctor is the one who's like . . . he's soaking up what I'm saying. He's the real story. The rest of them, they're just there for the show. . . . "

> "And who else was there?" I asked.

> "Well," he continued, "the one person who seems to be able to stand up to the doctor is this minor Shaykh who no one seemed to give a lot of credence to. But I find out his sector of the tribe is the most important. He just *chooses* to be very quiet and control things behind the scenes."

With the assistance of a local interlocutor, human behavior is convertible into information. Any opaque scene is penetrable.

·   ·   ·   ·   ·

Beyond these logics of optical and epistemic containability, human technology is also constructed through military ideas about making rapport with local individuals as an operationally instrumental process and a means of lightening one's own presence on the ground. These structures of feeling and imagination carry within them limit-points that are managed and policed by military personnel.

I spent a week attending intermediate regional studies classes with Special Forces soldiers in 2012, a class that was not part of the Special Forces Qualifying Course (SFQC), but created for those with higher foreign-language proficiency scores. The majority of students had recently graduated from the SFQC program. In the gray-walled classroom, the instructor announced to the soldiers: "By understanding a culture, you can manipulate people." The soldiers were to be trained in the PMESII-PT systems method (Political, Military, Economic, Social Infrastructure Information—Physical Environment and Time), a pedagogical tool also used more broadly by the conventional Army and the Reserves in their trainings.[14] Beyond mastering the PMESII-PT system, course goals included developing cultural competencies, evaluating one's own biases, and using cross-cultural skills to best accomplish the mission. The training soldiers were urged to cultivate an "open, outgoing attitude," and "obvious affection for the local people" and to speak "the native language of the region" in order to get "locals to seek you out" and "provide information on potential dangers." In a slide that began *Oakleys Off*, soldiers were further instructed: "Don't hide behind your ballistic glasses; Take off your sunglasses; Get out and walk around; Get to know the locals; Treat them as experts; Respect and Learn; Drink lots of tea; Get better every day."

The ideal native informant who will act as the soldier's entry point in the warzone[15] is a classic preoccupation constructed by Empire and the colonial gaze, a fantasy around penetration into otherwise secret or inaccessible zones.[16] In the contemporary retooling of the "native informant," soldiers are coached to forge "rapport" to produce operationally useful relationships with locals—i.e., native insiders—on the ground. Rapport is described as follows in the Special Operations Forces *Unconventional Warfare Field Manual*: "ARSOF personnel establish rapport with the local leadership by demonstrating an understanding of, a confidence in and a concern for the group and its cause. . . . Building rapport is a difficult and complicated process that is based largely on mutual trust, confidence, and understanding; it is rarely accomplished overnight" (ARSOF 2008, 4–8).

The conventional Army's guideline to counterinsurgency, *Field Manual 3-24*, is infused with a similar ethos, urging that: "[t]he interconnected, politico-military nature of insurgency and COIN requires immersion in the people and their lives to achieve victory" (Department of the Army 2006, 1–23). The manual contains an appendix on "Establishing Rapport," urging that: "Soldiers and Marines should establish rapport early and maintain it throughout the operation" with their interpreters and "should gain an interpreter's trust and confidence" (C-21). Meanwhile, the *U.S. Army Counterinsurgency Handbook* (Department of the Army 2007) urges the importance of cultivating rapport with all host nation interlocutors, noting that these individuals may help the soldier understand the social terrain: "[t]hough Soldiers and Marines should gain as much cultural information as possible before deploying, their interpreters can be valuable sources for filling gaps"—while cautioning that "information from interpreters will likely represent the views of the group to which they belong." Locals are thus mobilized to act as soldiers' entry-points into the warzone: meeting them at a precarious nexus of knowledge, affect, and risk.

Move like vapor, a captain had once yelled out to his training men, as I followed behind. He was quoting the tactics for the kind of war they waged from one of their models, T. E. Lawrence, the British soldier who had dressed in local garb, befriended the rebels, and served as their liaison officer during the Arab Revolt.[17] I had come to understand that one might "move like vapor" by land or by sea, leaving the most minimal trace, or during a training exercise—by not cracking a single branch. Yet as I observed the trainings, I increasingly became acquainted with the military protocol to move *culturally* like vapor as they cultivated rapport. In many field trainings (particularly in the case of the Special Forces) I witnessed soldiers were urged not only to *get to know* locals (to ask solicitously about their families, to chat about their villages, to comfort them in a time of sorrow) but also to grow their beards and to wear *kūfiyyah* scarves during deployments. This second imperative urged the soldiers to mimic the very bodily postures and expressions or even mirror the emotions of their interlocutors: in essence, to flirt with *going native*— becoming closer to cultural insiders by seeking to resemble those on the ground in the warzone, to understand/become them, and thereby gain power over them.[18]

.   .   .   .   .

A subset of crucial recent anthropological work on Empire has brought the lens of affect to bear on imperial projects.[19] Foremost among them, Ann Stoler examines the "emotional economy and the sensory regime in which these [colonial] relations were rendered possible" (2002, 165). At the height of the 2003–2011 Iraq War, affect was summoned, when Major General Robert Scales, one of the authors of the Counterinsurgency doctrine, wrote: "Understanding and empathy will be important weapons of war" (2006).

Yet within this call to empathy, there is an ongoing and uneasy tension between the soldier acting like a "surgical tool" of the state and cultivating an open attitude: they were urged to not "get too soft," and to use culture as a "means to an end." To this end, they were encouraged to go native but not *too* native: building of rapport ultimately entailed boundary work that had to be legislated. The making of rapport thus was ultimately *instrumental:* the soldier must be agile and light, become an insider only insofar as he or she has goals to accomplish. They must do this without becoming overly emotionally or imaginatively entangled with the circumstances and lives of others—their heavy weight and uneasy tethers.

"Empathy" in this case is thus tactical, must be deployed only up to a point, and may in fact be used to harm others (see Bubandt and Willerselv 2015; Stone 2018).[20] Soldiers are trained to strategically and "empathetically" mirror their wartime interlocutors, without truly seeking to understand those individuals' experiences (all while continuing to suspect them as potential insurgents or terrorists) in a kind of "imperial mimesis" (Stone 2018)—to gain power over those they mime.[21]

One afternoon, in a greasy spoon a few miles away miles from a mock Middle Eastern village where soldiers trained for deployment, "Carla," a soldier who assisted in the trainings, told me about cultivating "emotional connections" with one's wartime interlocutors, the "locals" on the ground. Carla said that she had learned to "connect" from a Human Intelligence soldier when she was deployed in Iraq: "He told me, if something bad happens downrange, you're allowed to hug the locals." Still, she explained that she had been bewildered as to when and how she was meant to engage in this manner.

Carla's lack of clarity over when one might console a local is but one example of boundary confusion. The Cultural Turn and counterinsurgency logics produced an affective quagmire for soldiers. In particular, at the beginning of this turn in the Iraq War, soldiers were urged to leave their guns outside locals' homes in a display of trust, or to even walk through markets unarmed.[22] Soldiers became increasingly baffled and alienated when these tactics ended up killing their comrades. Ultimately, evolving training exercises were structured around a dual principle: both a performance of openness alongside a hermeneutics of suspicion. Soldiers came to understand, through the exercises, that a bomb could be hidden anywhere: under a rug, in an animal carcass, under the dress of a pregnant woman. Indeed, in one training exercise I witnessed at a military academy, the trainees encountered a pregnant woman and learned that she was harboring a bomb. In a separate exercise later that week, they immediately assumed another pregnant role-player intended to blow them to bits, and thoroughly searched her, leading her to cry out that she had been humiliated. Thus, in the paradox of the Cultural Training, the training soldier is meant to open and ask their interlocutor to open, an act of mutual vulnerability and trust, but is simultaneously trained to assume the worst.

Amidst this complexity, predeployment trainings were designed as rehearsals of rapport, ways for soldiers to develop the right kinds of relationships in the warzone. After war game scenarios, psychologists ("Psychs") coached soldiers, in effect, on how to become cultural insiders via embodied empathetic performances. They were told how to craft their facial expressions and bodily stances, to be aware that they were being read by their interlocutors. One Psych told the training soldiers to be aware of their every bodily movement, all their "non-verbals": "Your whole person is being taken in. What does it say if you cross your legs versus stand on two feet," referring to a captain's stance in a recent exercise. The captain protested: "I was leaning in so I wouldn't be *above* [the role-player]. The Psych then arrived at his key point, urging the training soldiers to become mirrors of their interlocutors: "You need to find common ground, to humanize yourselves. Identify the norm of who you're talking to and try to *echo it*. Little tiny things like that: try to blend in with the guys you're talking to."

Popular media and public discourses have frequently lauded empathy as a balm for all social ills, as a bulwark against war, and as a means of holding society together (Hoffman 2000; Howe 2012; Krznaric 2015). Conversely, empathy and its variable embodiments and consequences are best put in context without a single value judgment (Holland and Throop

2011; Skultans 2008). Empathy can be understood as a kind of reasoning in which a person resonates emotionally with another's experience while imaginatively inhabiting that person's predicament (Halpern 2001). Meanwhile, the potentially darker and more deceptive qualities of empathy, rather than its idealized valences, invite further scrutiny: while it is often described as a virtue, empathy may in fact be used to harm others. For example, tactical empathy may help fashion the Other and even legitimize his or her destruction, such as a hunter dressing as the moose to inhabit the animal's position and to lure it (Bubandt and Willerslev 2015). Likewise, appointed soldiers prepare for war by growing their beards, wearing a *kūfiyyah* scarf, and miming the bodily postures of their Middle Eastern interlocutors so as to set them at ease and facilitate the extraction of information or the fulfillment of other operational goals.

Well-crafted mimetic performances rather than genuine experiences of empathy enable soldiers to know the Other to an extent that is useful to the wartime project: in this "Russian doll" model of empathy, it is possible to imaginatively understand the basic exterior circumstances of the Other (as in the outward layer of the doll) without experiencing undue concern for that Other (de Waal 2007). Similarly, war trainings unfold in a space where trainees are coached in careful attunement to the circumstances of the Other, but ideally not in *truly* caring for the Other.

One night after having barbecue at Jack's house in the suburbs of the military town, we sat outside swatting mosquitos, the pool fluorescent in the warm evening. Jack, a Green Beret in his thirties who assisted in crafting trainings, was eloquent, a thinker and a questioner wary of the status quo, and we had long meandering conversations about the United States, war, and power. When I asked him about cultural empathy, he explained that it was a master tool, crucial for the mission. Interestingly, he suggested that it is possible for the emotional content *behind* the presentation of empathy to be present or absent and for the surface to remain indistinguishable. "Culture is the game," he explained. "Having cultural empathy and having acute cultural understanding look no different. You can't smell the difference."

According to Jack, the masters of the craft could make empathy look identical to "honesty"—an honestly felt emotion or intention: "If I'm being manipulative and I want it to work, then I have to be so pure in my empathy that it mirrors in a perfect mirror, the honesty." Yet, according to Jack, this seamless indistinguishability was rare, and "usually more people can tell the difference. I can tell if you're trying to manipulate me."

The mosquitoes were skimming over the yard and pool and simmering around the porch lights, and I was slapping at them so much that we finally went inside. We sat down at the kitchen table, and Jack drew me an ascending pyramid in my notebook, dividing it into strata. He labeled the top tier the "Master Manipulators," to describe those who could manipulate and mime others with pure genius, absent any emotion, to achieve their desired end, and power over those they interacted with.

> Jack explained: "The manipulator goes beyond the system. His communication is like playing chess. He could even be mistaken for a native. The manipulators can disappear and come out on the other side; they can thread a needle and you don't know it."

> I watched, rapt, as Jack filled out the other sections, "And who is next?"

He described the next tier down as the "Elites"; those capable of both understanding the language (emotions and situation) of others, as well as speaking that language. The following category was the "Empathizers," those capable of perceiving a particular world and partially capable of communication: that is, either able to communicate or able to receive communication, but not adept at both. He labeled the next tier down the "Tourists," whom he described as those soldiers who were able to interact with others but not able to affect them and unable to interpret if their messages had been received. Jack explained that he had drawn a triangle to emphasize the excised portions of an original square: therein lay "the fat, those incapable of empathy." The Master Manipulator was thus capable of miming and coming to understand the Other without becoming emotionally entangled, to achieve an operational goal.

Jack's taxonomy of tactical empathy privileges an experience of imaginative reasoning unencumbered by feeling emotional resonance with an interlocutor. An ideal version of military empathy within the context of rapport making—"to thread a needle and you don't know it"—is much akin to what Danilyn Rutherford (2009, 4) calls "sympathy" within a colonial context, which refers to the affect generated when the colonizer eats, sleeps, and lives in the intimate company of the colonized while never giving up his place within the colonial hierarchy. These wartime affects "encompass empathy, pity, and compassion, but . . . can spawn hostility as easily as love." The encounter enables the soldier, ideally, to "re-create in ourselves the thoughts that must go on in the minds of others, especially those we dispossess" (Rutherford 2009, 4–5).[23] In an optimal military experience of tactical empathy, the soldier re-creates in his or her own mind the thoughts of the Other in such a way that does not interfere with ongoing structures of dispossession. In Jack's scheme, the more soldiers are able to mobilize an emotional and cultural acuity, the more agilely they might maneuver through that terrain. They must move softly and also smoothly, without becoming emotionally entangled. Entanglement is risky, it is knotted; it prevents a soldier from properly doing their job.

In order to avoid entanglement, emotional and imaginative identifications that might be established within the space of rapport were carefully policed by the military, so as not to create overflow. The soldier was meant to go *in* but not too far in—a seductive boundary zone that each had to learn how to negotiate. For example, Fred, a Vietnam-era Green Beret, described the capacity to connect as a soldier and to not only enter the world of the Other, but to *become Other*, and he emphasized the necessity of retaining a limit point:

> "You can turn it on and off. I was Korean. I was a Hashimiya Bedouin. You become more sympathetic. The mission is to live with them."

> "And do you feel you've actually become Korean or Bedouin?" I ask.

> "There are checks and balances," Fred replies. "The old guys [soldiers] make sure you don't go too far, don't become so sympathetic that you lose focus."

If a soldier goes entirely native, he crosses a threshold where he cannot "turn it on and off." "Jack," another Green Beret, invoked the dangers when a soldier couldn't "turn it off." He explained: "Cultural empathy can be a weakness or a strength. Some dudes go native. A certain kind of person would sacrifice himself for the [local] culture. But he needs to be pulled back." As was typical in many of my interviews with soldiers, Jack brought up Jim Gant as an exemplar of this tendency. Gant had been a pioneer, at the forefront of the U.S. military's "light footprint" doctrine, insisting that there be "an Afghan face on every mission" (2009, 6).

Many Special Forces soldiers spoke of Gant's tactics with wary admiration. Gant's book *One Tribe at a Time: A Strategy for Success in Afghanistan* details his "Tribal Engagement Strategy" to defeat the Taliban—to essentially become one of the local tribes. Gant begins his strategic manual: "Afghanistan. I feel like I was born there," explaining, "[m]y unit and I became family members with Malik Noorafzhal's tribe" (2009, 4).

According to Jack:

> "[Gant] was an original. He went into Southeast Afghanistan three times and got close with the warlords and tribal elders. He focused on culture as the operation, rather than on the operation itself."

> "And then what happens?" I replied.

> "You can go too far down the rabbit hole if you try to help the people too much."[24]

For Jack, if you focus too much on the people, you might lose sight of tactics. In his scheme for tactical empathy, adept soldiers will skirt the edge of that rabbit hole without falling down it: the more they mobilize an emotional and cultural acuity, the more agilely they might maneuver through the terrain of war. Enter the warzone, make the right relationships, find the right Others, and as you accomplish your goals, tread as softly as possible.

· · · · ·

One afternoon, I sit with Cal, a Special Forces major, on a lake on the rim of the training exercises. The grasses seethe with ticks, and we both wear a bug spray that is 100% DEET. Although I usually observe exercises, I have just returned from role-playing a guerilla fighter named "Gypsy" who was being trained by the (training) American soldiers, in an Unconventional Warfare rehearsal. The only woman present besides a cook who was not going out on missions, I had staggered several miles through the woods behind a darting team of soldiers, carrying enough water for twelve hours and other supplies on my back, asked to not crack any branches, to not trust anyone we encountered, and to leave no trace of my presence.

When Deleuze and Guattari describe the labor of becoming-imperceptible, they note that: "to go unnoticed is by no means easy." Yet perhaps it is easiest for the camouflage fish, which "forms a world" with everything that surrounds it: "It is crisscrossed by abstract lines that resemble nothing, that do not even follow its organic divisions; but thus, disorganized, disarticulated, it forms a world with the lines of a rock, with the sand, and with plants, becoming imperceptible" (1987, 279–80). Becoming-imperceptible requires ongoing *movement*, never fully landing, perhaps never fully being known as such.

Later when we're sitting at the edge of the lake, I ask Cal about trust in wartime. He tells me: "you act like you trust someone, but you only trust someone as far as you can throw them." Meanwhile, both the acquisition of trust and the success of a mission are correlated to *lightness* and *smallness* of presence—being less perceptible—a property he considers absent from the conventional forces, or "Big Army."

> Cal offers: "You go into a [Middle Eastern] village, and if there is a *large* U.S. presence like infantry, 'Big Army,' [conventional forces], they 'close with and destroy' the enemy. That's what they're trained to do."

> "And what are the outcomes?" I ask.

> Cal replies: "If you put them on an operation with intel[ligence], they have such *a big signature* that it tips off the enemy, and the enemy *evaporates:* they don't find anything in the village.[25]

According to the Cal, the soldier's relative invisibility directly impacts not only the locatability of the adversary, but the tolerance threshold and the willingness of the local population to trust the occupying American force:

> "The conventional guys get frustrated; they can't *find* the enemy. The second time [they enter the village], the locals get mad. You kick my door in, you know, I've had it with these guys."

He then contrasts the Big Army's heavy imprint to the movements and impacts of the Special Forces. These latter movements are imagined, conversely, as unseen and undisruptive, that is, much like the gliding of the pin amongst the pines:

> "But when *we* go into a village we're so *small*, we have such *a low signature.* We can *sneak in* and *pass ourselves off* as another vehicle, even if we're not sneaking through the cornfields with a knife in our teeth."

This logic of geospatial invisibility is next coupled with attempts to minimize presence through an ontology of chameleoning; that is, seeking to become insiders, via "culture."

> "We understand culture and we try to adhere to belief systems. It is less invasive, and the people are more accepting of us," he explains.
>
> "And where do the locals on the ground come in?" I ask.
>
> "In what we do, the Host Nation forces are the lead element that people see. There are Americans here too, but we've got beards, and we look like them as much as we can."

For the Special Forces, becoming an insider in the warzone is expressed not only through choreographed attempts at material, geospatial disappearance—that is, through secret night-time landings—but also through cultural disappearance, through "look[ing] like them as much as we can." Indeed, "[t]he real point of distinction for Green Berets compared to their door-kicking brethren is their ability to work by, with, and through foreign fighters" (Beech 2013). In the inward tumble's furthest expression, Green Berets seek to in fact disappear, displacing their own faces by acting *"by, with, and through"* local interlocutors on the ground as proxies.

Acting with, through, and via the assistance of local Others on the ground, while minimizing one's own traces, was a long-standing tenet of the Special Forces, but it arguably also became an essential, if more significantly more diffuse core of conventional military practices amidst ongoing conflicts in Iraq and Afghanistan. In 2012, a U.S. Department of Defense "Strategic Guidance" document laid out what is now known in policy circles as the Obama doctrine. The policies and programs it laid out were reflective of increasing preoccupations that had developed between 2008 and 2012, amidst a calculus of commuting violence elsewhere. President Obama stated: "Whenever possible, we will develop innovative, low-cost, and small-footprint approaches to achieve our security objectives relying on exercises, rotational presence, and advisory capabilities" (DOD 2012, 3).[26]

In this constellation, local Host Nation interlocutors would frequently be the ostensible agents of wartime action, while the American military advised behind the scene. This new way forward sought to diminish the most visible mechanisms and impacts of war, for the media and the American public as well as for American soldiers themselves. Meanwhile, *actual* American presence across the globe, most visibly in the form of the Afghan surge in 2009, but also in the pervasive form of Special Forces deployments, was increasing staggeringly. In *The Nation*'s January 2014 article, "America's Secret War in 134 Countries," Nick Turse documented how there had been a 123 percent increase in U.S. Special Forces deployments around the world during Obama's term, from deployment in sixty countries under President Bush to 134 countries under President Obama: "This presence—now, in nearly 70 percent of the world's nations—provides new evidence of the size and scope of a secret war being waged from Latin America to the backlands of Afghanistan, from training missions with African allies to information operations launched in cyberspace." More recently, President Trump dramatically increased the defense budget, and has expanded American military presence around the globe, extending Obama's footsteps and setting a record in particular for the use of Special Forces, deploying them to 149 countries in 2017—"75 percent of the nations on the planet."[27] Thus, "on any given day, about 8,000 special operators—from a command numbering roughly 70,000—are deployed in approximately 80 countries."[28] Moreover, amidst President Biden's withdrawal from Afghanistan, the administration has signaled that the U.S. military's global footprint will not otherwise shift in any major way at the moment.[29] In addition, in a reconfigured post–Cold War American Empire, the U.S. has dismantled its larger military bases while constructing tiny and less visible "lily-pad" bases across the globe. In some cases, these bases are even tucked within a host nation's base, thereby seemingly effaced (Vine 2015).[30] Much like this logic, the mandate of the Special Forces also renders U.S. wartime presence abroad less visible. In unconventional or guerilla warfare, all acts on the ground must be accomplished covertly "by, with, and through" a local member of the host or occupied nation. These proxies are perhaps human technologies par excellence, performing local wartime labors for the U.S. military, the originary agent of those acts.

Although not accomplished through covert acts, this logic has increasingly permeated the conventional Army as well. In 2013, a soldier explaining the idea of the smaller footprint described to me the way his presence was diffused within an Afghan village: "Usually, the provincial governor would hang up a picture of the American commander who was there at the time, next to his own, in his office. I told him, no, just hang up your own. It's only you here." The U.S. soldier's refusal to hang up his own picture in the Afghan provincial governor's office points towards an unsettling American military trend to at least partially mask military presence, commuting experiential and lived costs elsewhere.

Back in the mock villages of the Special Forces, training soldiers learned these lessons most emphatically. In a basic counterinsurgency/culture exercise, the soldiers enter a collapsible house with prayer rugs adorning the walls and a bowl of fake bananas and grapes on the table. Muhsin, role-playing a "Key Leader," a red-checkered *kūfiyyah* wrapped around his head, pours real tea into a cup for the soldier. They must drink together if they are to seal an agreement. With the help of his interpreter, the soldier asks Muhsin about the state of the village in order to fill in a map of the "human terrain." Through the course of the conversation, Muhsin tells the soldiers about the insurgent elements in the village. In the After Action Review, the instructor tells the training soldiers to curate the action into the hands of the locals and so that they stay more backstage: "Your primary objective is not to be Mr. Fix-It: your U.S. Type-A change how things are [attitude] to how they ought to be. Don't synchronize the village: let the [Key Leader] do it!"

Yet the costs on the ground for the local individuals enlisted by U.S. forces are typically far more grave than for the Americans behind the scenes. The camouflage fish seamlessly and elegantly "forms a world" with everything that surrounds it. However, within the human social, this form of disappearing-into-everybody demands craft: "This requires much asceticism, much sobriety, much creative involution: an English elegance, an English fabric, blend in with the walls, eliminate the too-perceived, the too-much-to-be-perceived. Eliminate all that is waste, death, and

superfluity, complaint and grievance, unsatisfied desire, defense or plead-
ing, everything that roots each of us (everybody) in ourselves, in our
molarity" (Deleuze and Guattari 1987, 279). What they are describing
here is the shedding of that which imprints particularity: mannerisms,
identity markers, "everything that roots each of us . . . in ourselves." This
releasing of markers requires movement and sheer speed.

In Deleuze and Guattari's formation, this releasing is potentially mysti-
cal: "The imperceptible is the immanent end of becoming, its cosmic for-
mula" (1987, 279). However, when we re-enframe the act within a political
space, indeed within the space of *war*, the ethics of becoming-imperceptible
become pertinent. The warscape kills and lets live, befriends and reviles,
creates friendships and betrays, based upon particular identity. In war, the
American soldiers who seek to become imperceptible try to become second-
ary behind the faces of nationals—they covertly train local guerillas to overthrow their own governments; they don't hang their picture in the Afghan
provincial governor's office; they ask local individuals to take the risk of
revealing information that could get them killed. Meanwhile, those individuals become newly and permanently marked within their own countries,
differently conspicuous and endangered.

Human technology is predicated on a military dream of extended visibility, agency, and lightness—becoming insiders—where the so-called
local Other ideally bears the greater cost. This paradigm ultimately
depends on a presumption of the knowability, maneuverability, and containability of the world outside the self. Human behavior, patterns, intentions, wishes, dreams, longings, contradictions, might ultimately be
mapped; collateral human damage may be excused away within a bigger
project to buttress Western liberal imperialism.

[Field Poem Fragment]

One soldier
who roasted a pig on his porch   barbecuing until sinews were tender

tells me he waited above the Euphrates waters and if they tried to pass
even after we told them not to, they deserved it: pop (deserve it); pop

(deserve it). Euphrates, your dark tunnel out is rippling around us.
In the war, a child approaches a tank as one soldier counts the child's

steps. In the town, I drink a bottle of wine with that soldier
among barber shops, boot repair shops. Is she my friend? I weep to her.

I've lost who I thought I loved and she says I did
this thing and to whom was that child beloved?

Find common ground, the soldiers say. Humanize
yourselves. Classify the norm of who you're talking to, try

to echo it. Do this for your country, says one soldier; we
are sharks wearing suits of skin. Zip up.

2 The Iraq Warscape and the Cultural Turn

It was an arid day, and we could hear distant, tinny gunshots—all blanks—from all directions on the military base. I sat with Mohammad, an Iraqi role-player, and Joe and Bill, two American soldiers who had recently returned from a deployment in Iraq. We were among a bigger group of soldiers.[1]

"Did you ever get to know Iraqis when you were in Iraq?" I asked the soldiers. "Did you work with them?"

"Terps" (interpreters), answered Joe.

"I got to know a couple. Yeah, they were great," offered Bill.

"They were gay?" burst out Joe.

"Great."

"I left my shoes outside one day, and the next day they were gone," added a third soldier. He did not need to say that he believed the interpreter had stolen them. The soldiers began laughing.

"Man, and they were brand new shoes. I bought them two days before that. They were gone."

"Actually," added Joe in a funny tone, "I found my terp dead. He was beheaded outside."

"Oh God," I blurted.

"Yeah," offered Joe, letting out an uneasy laugh, his Adam's apple bob-bing in his throat. "It sucks."

We sat in the unflinching sunshine, the fake guns going off, *pop, pop, pop*. Everyone was quiet for a moment. I watched Mohammad, the Iraqi role-player, but he evaded my attempt at eye contact. Joe tells me he only knew his interpreter a bit before he was killed: "We talked once in a while. Cause he told us [the militia] killed his family first. He was telling us about that. And two days after that, we found him beheaded." The conver-sation then turned to the circumstances of Iraqi interpreters during war-time. Joe explained that his interpreter had left the base and gone home a lot, risking his life. When I commented that the life of a war interpreter was a dangerous one, another soldier chimed in, saying that Iraqi inter-preters received death threats "like four times a day." Another soldier then abruptly offered:

"There was a terp who turned out to be working for Al-Qa'ida. It was a chick."

"Oh, we had one of those," added yet another.

"Did you trust your interpreters?" I asked.

"I really don't trust terps," said Joe. "You gotta learn, if they're around you enough. But I don't trust them."

"You can start trusting them when they kill one of their own. I've seen them, and that was pretty cool," said Bill. "You start trusting them then."

.   .   .   .   .

During the American occupation of Iraq—a war with a death toll of 182,000 Iraqis, and three and a half to five million displaced—a complex landscape of alliance and affiliation emerged.[2] Somewhere between 30,000 and 100,000 Iraqis worked directly for the U.S. military, govern-ment, media, contractors, and subcontractors during the 2003–2011 Iraq War as interpreters, translators, contractors, fixers, and bodyguards (Ondiak and Katulis 2009, 4). Many more Iraqis assisted the Americans informally by giving them tips, or worked with them as government func-tionaries to stabilize their districts and villages. Not unlike any other war in human history, the occupying U.S. military in Iraq sought to make sense of and maneuver effectively within the local terrain with the assistance of local interlocutors. The soldiers had been trained to try to find out "the ground truth"—Army slang for the reality in the warzone, which seemed to be most immediately available via those who were on the ground. However, an idealized American military construct central to the Cultural Turn—employing local intermediaries as repositories of knowledge, a mechanism to become insiders in the warzone—played out in charged and unpredictable ways during the 2003–2011 Iraq War. Even after a militia executed Joe's Iraqi interpreter for his affiliation with the American military, Joe's impulse was not to note the peril interpreters were put in or the sacrifices these individuals had made to help American soldiers on the ground. Conversely, in the aftermath of this event, he noted that he did not trust Iraqi interpreters as a whole. Meanwhile, Bill believed that he could only fully trust an Iraqi if he or she expressed allegiance to the American troops by turning against and killing one of their countrymen.

American military personnel imagined wartime locals on the ground as potential human technologies and so turned these individuals into repositories of culture, within a model of business and war that confirmed their expendability. Meanwhile, amidst an American military dream of transparent oversight and omniscience in wartime, the warscape itself was full of indecipherability, fear, and mistrust. This indecipherability was a political product of confluence of factors: the aftermath of Saddam Hussein's totalitarian regime, the American occupation of the country (and the new intelligence and reconnaissance technologies that accompanied it), a complex landscape of insurgency and counterinsurgency and an influx of money from outside Iraq's borders. In this environment, friends and enemies are not just policed but are produced through wartime imagination (akin to "the state's work of creating human beings, of 'making up people' as Ian Hacking puts it" [Verdery 2018, 29]). It is in just such a wartime milieu where the "structures of recognition and speaking on behalf of others" can become readily "undone" (Mojaddedi 2019, 518) amidst war's condition of "nontranslatability" (Apter 2006, 19).[3]

The work of the translator is intrinsically complex[4] and fraught with power.[5] Recent thinkers have described the work of translation as that of crossing a gulf and thus subject to gaps and ruptures—a "theoretical metaphor through which to think about difference" (Giordano 2014, 15), and "making it possible to wage war on those who were previously kindred" (Mojaddedi 2018, 501). The translators in this particular story were regularly summoned into the work of cultural translation in an intricate web of power relations. Indeed, the metaphor of cultural translation itself is intrinsically violent, "enmeshed in the conditions of power" and generally occurring unidirectionally, from East to West, to be consumed by Western readers (Asad 1986). Positioning the work of translation in wartime amplifies this complexity and the possibility for rupture. Cultural intermediaries and translators have been fantasized as offering a panacea of mutual understanding in wartime; conversely, translation might be understood not as "antidote to war, but one of its central practices" (Colla 2015, 4).[6]

After the United States invaded and occupied Iraq, although moments of mediation and cultural translation and communication occurred, they did not typically do so within idealized or anticipated terms and often came with great costs. While some U.S. soldiers I interviewed genuinely relied closely on their Iraqi intermediaries, they just as frequently ignored Iraqi intermediaries' advice or accused them of having agendas, doubling for the other side, or doing the work only for the money—effectively producing their expendability within the American imperial project.

Meanwhile, as conditions in Iraq deteriorated and U.S. military promises remained unfulfilled, Iraqis who signed up to work for the Americans were likewise not trusted by their own countrymen. Frequently accused of betrayal by others (both Americans and Iraqis) and inhabiting a dangerous and ambiguous position, they had to maneuver carefully within the warzone's already unstable field. One by-product of the negotiation of their wartime choices and the instability of the field they navigated was, in some cases, wartime performance. As a multiplicity of shifting perceptions (with life-and-death stakes) began to cohere around Iraqi intermediaries, they were increasingly compelled to publicly perform a panoply of affiliations, roles, and affects, according to circumstance—for example, Iraqiness, Sunniness, Shianess, Westernness, earnestness, indifference, and so on—in order to negotiate the emerging and shifting political landscape, and in some cases, to survive the war. Mobilizing and demobilizing identity as necessary, these individuals lived contradictory choices *without* being absent allegiances (Kanaaneh 2008).[7] Indeed, perhaps the production of multiple forms of identities and even "doubles" of the self is key feature to living under surveillance and suspicion (Verdery 2018). These moments of more curated performance of affiliations coexisted with "the rough ground of mundane affairs and encounters," a space of more complex and daily entanglement with others that defied those reductions of identity (Al-Mohammad 2012, 44–49).[8]

As I listened to Iraqis in diaspora narrate their memories of working with the American military during the Iraq War, I traveled between brocaded couches and impoverished neighborhoods in Amman, government-subsidized housing in Louisiana to long walks in Central Park, and winding conversations at Manhattan bakeries to making biryani with Iraqis in North Carolina. Across these contexts—even amidst class, occupation, and regional differences—these individuals told of the (in some cases life-saving) importance of performance in some dangerous instances in the warzone, as they were read differently by different social actors.

Let me be clear: the absence of trust in these public and dangerous spaces—certainly between Americans and Iraqis, but also between some Iraqis and Iraqis they did not know—was historical and political, variegated and complex: a product of the impress of power and its shifts. The social interfaces of Iraqis (and of Iraqis working for the U.S. military) were not in any way uniform or static—and certainly in no way originary or *cultural* as the U.S. military might imagine it. Indeed, they were deeply influenced and largely produced by the politics and violences of American occupation itself, the most recent iteration in a long imperial history. The lifeworlds of these individuals could in no way be compressed into cultural/ontological bullet-points in a military handbook, a military intent on telling a primordial story that effaced American occupying presence itself and the complexities of politics and history.

. . . . . .

Remembering the spring of 2003, American soldiers I spoke with frequently described Iraq as being in some way illegible to them. Sometimes they described how the weather—what one soldier I spoke with called "moon dust"—blinded them: "After a dust storm, you can taste it. It gets in your weapon, in your vehicle. It collects on grease and makes it gritty. You can't see anything. The guys said it looked like it came from the moon." Meanwhile, Saddam Hussein had burned the oil fields, which reduced visibility. A former Iraqi interpreter describes the scene to me: "Baghdad was covered with this huge big thick cloud of smoke, black smoke. The trees were black, everything was black, black, black." As military personnel tried to engage with the Iraqi population—equipped largely with their own orientalist projections, and refusing to see their own presence itself as a meaning-making gesture—they described Iraq as increasingly illusive and obscure. These discourses of the unfathomability of Iraq are a classical colonial trope across time and place, as well as specifically echoing British colonial discourses.[9] In the American media and even among many academic writings since 2003, Iraq has likewise been described as illegible, unknowable, violent, disorderly, indeed "ungovernable" (Dewachi 2017). In the paradox of Empire, such discourses misrecognize such violence as originary to Iraq, rather than produced by Empire itself.

The American troops entered Iraq in 2003 with minimal knowledge about Middle Eastern culture, typically perceiving culture as "irrelevant" to mission effectiveness and cultural training as "useless." Indeed, at the outset of the war, the military "had a scattershot approach to cultural training—recycling old materials and hiring contractors to churn out handbooks, compact discs, and Power Point presentations about Iraq, Arabs, and Islam" (R. Davis 2010, 9). Between 2003 and 2006, over 1.8 million "Iraq Culture Smart Cards" were distributed to American soldiers, offering bullet-point synopses of Iraqi and Islamic customs, gestures, and clothing. Yet the content was often more telling about U.S. military culture than Iraqi culture: "For one thing, the strict order of meanings assigned to various types of headdress parallels the Uniform Explanation Chart of the Marine Corps, which determines who should wear what color uniform when" (R. Davis 2010, 10).

Upon entering Iraq, soldiers often found themselves unprepared to comprehend the country's complexities. One former soldier described to me the complexity of the social groups they encountered: "Baghdad is the size of Chicago, and it has that Chicago feeling to it. There was the former government. There was the black market. The people running on the street. It was very difficult to identify who is who. We needed intermediaries."

Amidst a social landscape U.S. soldiers deemed indecipherable, Iraqi intermediaries became central to the American military project. In the earliest months of the war, Iraqi intermediaries were drawn from the "Free Iraqi Forces," Iraqi expatriate (American and European) opposition volunteers typically linked to Ahmed Chalabi's Iraqi National Congress. As the war progressed, the need for more Iraqi interpreters and other contractors became dire. Several Iraqi interpreters told me that if you had basic English conversational skills in this period, you could be hired off the street—and in the military dream, then explain that street so it could be subdued.

·   ·   ·   ·   ·

In the early years of the 2003–2011 Iraq War, a confluence of complex processes unfolded. An influx of Iraqi exiles who had escaped Saddam Hussein's regime returned. Members of the Iraqi opposition, some remaining abroad, came to have the ear of the architects of the U.S. occupation, calling for a "liberation."[10] Vendettas emerged against prior members of the Ba'ath Party. A new militia culture was emerging, with affiliations to Iran. In this brew of shifting politics and collaborations, U.S.-affiliated Iraqis likewise emerged. Choosing to work for American forces and companies was likewise shot through with complexity. Many of those I interviewed described both an initial optimism about disposing of a corrupt regime and urgent economic motivations to feed their families in a troubled time. The translators constitute one micro-world in this war, subject to a tumult of larger forces and politics, especially as resentment grew among the Iraqi population against the U.S. occupation. Many of these individuals narrated their own positions as becoming more fragile over time, as initial feelings of possibility and promise after Saddam's fall soured into disillusionment.[11]

"Ahmed," who became a volunteer interpreter the first month of the war, reminisced to me about early enthusiasm and hope.

"The first week or after two weeks, after the Occupation, the Iraqi families, they used to serve [the troops] breakfast, lunch, *and* dinner," he explained.

"What was that like?"

"You know, they start from the windows, youyouyouyouyou [ululation]. A lot of people came with big trays with sherbet, chocolates. The whole street stood with [the troops] and took pictures. This is not a lie. When they said in the news, they offered them roses—they offered them everything!"

Yet this brief window of hope, for those who felt it, soon began to narrow and then shut. After the collapse of Saddam Hussein's rule just weeks after the beginning of American military strikes—and the spectacle of toppling Saddam's statues—President Bush declared "mission accomplished" in May 2003. In August of the same year, a suicide bomber blew up a portion of the United Nations compound in Baghdad, incurring heavy Iraqi casualties, including killing Sergio Vieira de Mello, the U.N. secretary general's special representative to Iraq. As one of the key agencies charged with rebuilding the country burned, it became clear that the war was hardly over. In March 2004, after four Blackwater security employees were killed and mutilated, American forces launched the First Battle of Fallujah, causing much of the city's inhabitants to flee, catalyzing broader fighting across the country, and amplifying Iraqi resistance to the war. In this period, popular Iraqi opinion turned decisively against American occupation, ossifying as photographs of U.S. soldiers torturing and mocking Iraqi prisoners at Abu Ghraib began to circulate.[12]

As Iraq was increasingly destroyed—both by U.S. occupation and by increasing and various militia presences—the Iraqis who signed up to work for the Americans were more frequently accused of betrayal by their countrymen. Yet if the traitor "holds up a mirror to society of all that it most fears about itself"—exposing the potential "contingency" and "frailty" of our commitments (Kelly and Thiranagama 2010, 10), these Iraqi individuals must be understood otherwise.[13] Hardly absent ideological commitments, they chose to work for the U.S. military for a panoply of ethical, political, and economic reasons, which acquired more contradictions as the war progressed. Even for the many who believed the war was an imperial invasion, it was also initiated against a regime that the population abhorred. Meanwhile, the return of opposition members from exile who were already working with the United States—yet another collaboration—added complexity to the question of affiliation. From the outset, the contested legitimacy of the regime rendered the question of collaboration ambiguous. Rather, these people must be understood as "conscripts of modernity"—inhabiting and making choices within conditions not of their own making[14]—rather than "native informants" as such.[15] They are also crucially translators—and like wartime translators across all of history, desperately working across difference—its ruptures, its gaps, and moments of total incommensurability, and trying to live amidst the unlivable conditions of war.

One Iraqi former interpreter to the American military explained to me:

> "In the very beginning of the war, they only gave us five dollars a day. So you can see that those who signed up and worked for so little in that time really cared about the cause."

> "And how about those who signed up later?"

> "Six months into the war, they were already given $50 a day, and after that, more, so people who cared most about money joined."

As political circumstances shifted around them, these individuals inhabited increasingly complex positions. Jean-Paul Sartre proposes that the collaborator suffers from "historicity," and "ratifies events simply because they occur," seeking simply to side with the victor (2008, 57). However, Iraqis employed by the American military lived a much more morally complex form of history and temporality. Most often, their wartime choices were motivated in variable terms by hope for a stable future.

Those who signed up did so for a tangle of ideological and economic reasons, alternately describing their dream of bettering their countries and their own family's stability, while trying to negotiate shifting politics and threats. Others had enjoyed class privileges growing up and had access to resources outside Iraq, producing affiliations with the West. Some of my interlocutors imagined nationalism as a more moral motivation to work than financial gain. Supporting one's family in a troubled time was also driven by a moral urgency for many. Across these many variable cases, translation and other contracting work provided wartime *sustenance*—a way to survive and support and feed one's family in a time when survival was in no way guaranteed.

In the 1990s, Marwan was an ambitious kid with an itch towards the outside. One day, a Western man in a suit walking out of the U.N. building approached the child on the street and asked if he was interested in his future. Marwan knew only the words *hello* and *goodbye* in English; he dashed home to his father and asked what the word *future* meant. The next morning, he ran back to the same neighborhood and exclaimed: "Future! Future!" Marwan began running errands for this man, shining shoes and other odd jobs. During the economic hardship of the American sanctions of Iraq, Marwan became the sole breadwinner for his parents and many siblings. His neighbors and friends called him *ibn ajnabī* — foreign boy. In 2003, he immediately took the exam to be an interpreter for the U.S. military.

Like Marwan, the very first wave of Iraqis who worked with the Americans was often particular. Many of them had always felt a lure and connection to the outside: a passionate young woman interested in translating English literature; a young man who had lived in London while his parents had gone to medical school, returned to Iraq as a teenager, and was a junkie for Hollywood movies. Some had a genealogy of relationships with foreigners, their parents having worked with the British during their own childhoods. A great many of them felt their ambitions marginalized by Saddam Hussein's totalitarian regime and the requirement to be a compliant member of the Ba'ath Party in order to earn an exemplary score on the country's college exam. Many, especially in the first wave, were middle class and college-educated, but not necessarily extensively trained in foreign languages. For many, working with the Americans was an opportunity to practice English and to forge links with the outside. They constitute one point in a vast history of translators and contractors in wartime around the globe—one story of a group of people and their complex motivations.

In his 2007 article *New Yorker* article "Betrayed," George Packer describes the uniqueness and exilic marginality of this group of Iraqis: "Before the war, their only chance at a normal life was to flee the country— a nearly impossible feat . . . I thought of them as oddballs, like misunderstood high school students whose isolation ends when they go off the college" (2007).One young activist, described her motivations to work with an American company, with the hope of furthering a civil society change even though she did not believe in the American military intervention. Salam, who had lived abroad as a child while his parents went to medical school, and who worked as a translator for the Americans training Iraqi police, described how he decided to work for the Americans:

"We were very excited at the beginning. I applied to every organization I could find: the Red Cross, the United Nations, anything you could imagine, then I got this job with the Americans."

"And what did this job mean to you?"

"The role of a true interpreter is like bridging, to build trust and bring out the best in both sides."

Mohammad, the son of diplomats who had also lived overseas as a child and returned to his extended family's village right before the American invasion, described how he became an interpreter:

"I was probably the only English speaker in the town. It was a small town, like 2,000 people. About 70 percent of them never went to school. And most of them didn't study any English."

"So it fell on you, then?"

"Yes, so the same day the troops got there, I went down and spoke with them and started translating."

Although a number of the Iraqis interviewed described existential dislocation under Saddam's regime, just as many were focused rather on economic gain and desperate survival: a cab driver who became a driver for an Iraqi company working for an American company; a guard who worked in front of the U.N. because his aunt worked in the cafeteria in the same building and the family desperately needed the money. Sayf, who worked for American contractors from 2003 through 2009, told me: "If I find work, I work. The best paying work at that time was with the foreigners." 'Abd al-Qadir, a war amputee, retorted with indignation when I asked his motivation to work as an interpreter: "Economic." In many cases, Iraqis worked with Iraqi subcontractors and had no direct link to the Americans. Hussein, a mechanic during Saddam's regime, worked as a truck driver for an Iraqi company affiliated with the American military. He explained his choice of work: "I did it because I needed to feed my family. I never saw the American military or the American contractors. I dealt with the Iraqis who worked for the contractors."

Still others felt crushing internal contradictions: although they were against the American occupation, they had still chosen to work for the

U.S. forces. Whether they did the work out of economic necessity or a desire for the betterment of their country or a combination thereof, the choice produced an ongoing existential toll. One former interpreter, Mohammad, described this phenomenon to me: "What baffled me was how much self-hate some interpreters have. They talk shit about the U.S. Army while they're working with the U.S. Army."

Mohammad asked another former interpreter, Osama, about this challenge: "I was asking him, will you be able, ten years from now, to be 'out' about working with the American army? He said 'No.' He thinks that it's okay to be closeted about that. He thinks that people have the right to not understand that." According to Mohammad, this meant that people were "living a lie." Others described the condition as "living a contradiction" that was produced by wartime circumstances.

In each instance, these Iraqis, regardless of their motivations for working with the Americans, suffered devastating costs, and made a wartime choice that rendered them expendable—cared for inadequately by the Empire that employed them. A former interpreter who justified her work through an ethics of bettering her country, was threatened by a militia. Salam and Mohammad, who both sought to mediate between the Americans and their own people, each fled Iraq after multiple death threats. Hussein, who worked as a truck driver with an Iraqi company that supplied the American military, "did not even know who [he] was working for" and never once spoke to an American, watched as his nine-year-old child was killed in front of him in retribution.

·   ·   ·   ·   ·

The gray zone of the Iraq warscape—wherein "the 'we' lost its limits, the contenders were not two, one could not discern a single frontier, but rather many confused, innumerable"—was a space of indecipherability rather than of clearly delineated or articulable allegiances to nations, causes, or groups (Levi 1988, 27)—a space where one could not easily read or trust one's acquaintances or speak freely in the public realm.[16] The contact points between totalitarian and imperial, liberal logics offer an entry point into this period.

Saddam Hussein's regime, in which all citizens were asked to identify only with his person and the nation and to inform on traitors, cultivated both apparent uniformity and heightened opacity. In 1979, Saddam Hussein enacted a dramatic purge of the Ba'ath Party after announcing a narrowly thwarted plot on his life. One-quarter of his Revolutionary Command Council was executed and the nation was asked for vigilance: "A special telephone number, supposedly his own, was flashed on television screens to be used by informers wishing to squeal on 'enemies of the revolution'" (Aburish 2002, 55).[17] In police states, public participation in policing is cultivated and suspicion is the necessary condition, wherein the population was viewed both as a potential threat and as those to be protected. The police appeared as both part of the local community and apart from it (Feldman 2015).[18] Indeed, amidst such mistrust and potential surveillance "every conversation with someone becomes anchored by the presence of a third, often hidden" (Verdery 2018, 19).[19] According to some estimates, one-fifth of the Iraqi labor force was involved in the party militia, army, and police, and an unknown number of the general population acted as informants (Al-Khalil [Makiya Kanan] 1989, 38). Little has been written in English on this period, and the sources are muddy and ideological: for example, the writings cited by Kanan Makiya, one of the architects of the U.S. occupation, for example were key to producing Iraq as a country requiring liberation. Meanwhile, Ali al-Wardi, an 1950s Iraqi sociologist (who studied in the U.S. and was influenced by Ruth Benedict's culture and personality work), described the "schism of the Iraqi personality"—in essays that would be translated and co-opted by the U.S. Naval Institute in 2013 (Dewachi 2017, chap. 5, n9). That is, certain experiential conditions of this historical period were *both* lived *and* reified from the outside, infiltrating local discourses.

Mona, who worked as an interpreter during the 2003–2011 Iraq War, described the dichotomy of Saddam's regime:

"You had to hide your intentions, your real intentions, and you cannot express yourself, what you think, what you believe."

She explained that because of this period, Iraqis are habituated: "Iraqis are used to manipulation—the double life goes back to Saddam's time."

In totalitarian settings, this idea of a double life keeps the notion of an interior largely undisrupted and intact, as citizens enact compliance without necessarily believing in the spectacle that is produced. While Hannah Arendt (1976) proposes that psyches are colonized under such regimes, the situation of the Iraqi interpreters I worked with more closely follows Lisa Wedeen's reading of the interior life under totalitarianism.[20] Wedeen proposes that citizens may instead inhabit a state of self-aware complicity, acting "as if" they believed, while privately holding other beliefs (1999, 76).[21] That is, the totalitarian regime may create the conditions for masking—the motivations of others may be opaque and indeterminate, even while other beliefs might be held privately.

Iraqis I interviewed about this period described such an ongoing situation of duality. One man, the son of an ambassador, described his confusion as a child about the disparity between his father's public and private pronouncements about the government:

> "On the one hand my dad defended it in public and would give the cliché answers. And then he would go back in, and when I would hear him discussing these topics with my mom, it was something completely different."
>
> "And so what did you conclude?"
>
> "I was very confused—which is right, which is wrong?"

Meanwhile, this doubleness manifested in a feeling of prohibition against discussing the very images which surrounded the public: "At the same time, we were not allowed to bring it up at all, we were not allowed to bring up Saddam's name. Although I had his pictures everywhere growing up." Others described how people just wanted to get by, even if they didn't believe in the rhetoric of the regime. One woman, who later worked as an interpreter, explained: "You had to hide your real intentions. You can't express yourself. All your affiliations should go directly to the one party."

When the American forces invaded Iraq in 2003 and a new complex gamut of politics was produced—from the presence of opposition groups returning from exile in London to those affiliating with militias linked to Iran—that more monolithic performance shattered into a multiplicity of shifting performances for Iraqis who chose to work for the U.S. military. As popular sentiment against the war grew, those individuals soon had to conceal their work from their neighbors and acquaintances, and even in some cases from their families, to protect them from anyone who might to be linked to a militia. Meanwhile, even if they strongly supported the American cause, these Iraqis were still often imagined by the U.S. soldiers as potential spies, and thus had to continuously demonstrate their adherence to America. Indeed, Wedeen's "as if" model is too uniform in the Iraq context to describe the panoply of performances and underlying commitments that occurred during the war.

Verdery reminds us that "we are all multiplied by those we meet"—a universal process—but one that is intensified "when major cultural boundaries are crossed" (2018, 24). While "the performance of everyday life" (Goffman 1959) renders the texture of even the quotidian dramaturgical, in wartime contexts, sustaining particular impressions of the self and/or the self's affiliations have life and death stakes. For Erving Goffman, individuals labor to control aspects of their presentation so that the others in the social circle might accept their "definitional claims"—both of themselves and of the situation, and ensuring that these claims are not interrupted (1959, 10).[22] In the instance of Iraqis who chose to work with the U.S. military, different projections (pro-Iraqi; pro-American; pro–particular militia; etc.) were sustained in changing contexts as a matter of survival.

Meanwhile, this ethos of masking became charged with the accusation of having no ideological commitments at all, or "nothing" underneath (Kelly and Thiranagama 2010).[23] In this paradigm, pure economic motives were often read as an *absence* or *negation of* national, religious, or ethical commitments. Both American soldiers and, in some instances, other Iraqis suspected Iraqi intermediaries of working for the Americans for mere gain, and thus being willing to switch to the highest bidder. Conversely, my Iraqi interlocutors typically described their wartime choices as motivated by prior, deeply felt structures of affiliations or beliefs—driven by a desire to aid their country and feed their family. However, they in some instances imagined *others* doing similar work as driven "only by the money"—implying that financial gain was a less moral or less legitimate motivation for doing the work than nationalism.[24]

One former interpreter explained: "Most of the others only did it for the money. In that time, a lot of people would do any kind of work for money."

Another former interpreter, after explaining that many of the Iraqis who worked for the U.S. military did it only for the money, immediately segued into a discussion of darker wartime acts motivated by money:

"Many of the others who worked for the U.S. military did it only for economic reasons. People would do most anything for money. You want a sandwich—someone offers you $500 to shoot a U.S. Humvee."

"And then what?"

"That money is bad. If you use that money and you buy a sandwich, that sandwich is poison."

The wife of a former contractor with the American military echoed the sentiment: "Everything was about the money. No one believed in anything." In this imagined wartime mode of being in the world, a person is willing to do anything for the money, switching sides when necessary, while simultaneously not believing in anything. This masked figure was never a self-description, but rather, something of a phantom—a fear that others lived this way.

The terms of this masking, and its phantoms, were forged during the economic shifts of the Iraq warscape, which was often constituted by circuits of money, rather than by clear allegiances to nations or groups, generating indecipherability for both Americans and Iraqis. Logics of ambiguity under Saddam Hussein's regime were compounded by these shifts. During the Iraq War, crisscrossing and sometimes obscured circuits of money moved into the hands of both hired mercenaries and everyday people desperate for cash. In the immediate run-up to the war, former defense secretary Donald Rumsfeld emphasized the importance of the private sector, declaring in an article in *Foreign Affairs:* "We must promote a more entrepreneurial approach: one that encourages people to be more proactive, not reactive, and to behave less like bureaucrats, and more like venture capitalists" (2002, 29).

Experts argue that the 2003–2011 Iraq War was the most privatized war in U.S. history: during the Gulf War, there was one mercenary for every one hundred soldiers; at the beginning of the 2003 Iraq invasion, there was one mercenary for every ten soldiers; and by 2007, there was one mercenary for every 1.4 soldiers (Klein 2007, 380).[25] This analysis is perhaps overly stark, and seeming to minimize the presence of nearly two million American "troop-years" of American presence on the ground between Iraq and Afghanistan by 2013.[26] Nonetheless, privatization and a lack of accountability of civilian defense staff was a central trend during the war. Indeed, private security firms and defense contractors faced no legal consequences or oversight during the first five years of the American occupation of Iraq (Scahill 2007, 9). Immune from regulation, scamming was endemic to the period: "The big contractors engaged in elaborate sub-contracting schemes. They set up offices in the Green Zone, or even Kuwait City and Amman, then subcontracted to Kuwaiti companies, who subcontracted to Saudis, who, when the security situation got too rough, finally subcontracted to Iraqi firms, often from Kurdistan, for a fraction of what the contracts were worth" (Klein 2007, 356).

In this environment, economically driven social actors were sometimes several degrees removed from those who hired them, often making them difficult to read, and the contracting companies themselves were opaque.[27] Many of my interlocutors described such experiences of ambiguity in the warzone, for example how Iraqi subcontractors worked for American companies without ever crossing paths with their "parent" employers. Furthermore, rumors circulated among both American personnel and the Iraqi population about how in some instances, Iraqis affiliated with militias also signed up to work as translators and were paid by both sides. While actual cases of "turning" were infinitesimally small according to my inter-locutors, the uneasy imaginary of this doubling took on a heft in social life. Meanwhile, I was told that Iraqi government officials forged alliances and took bribes from militias, promising to protect them for votes. And, fur-thermore, as the Iraq border became increasingly porous, foreign militants, mercenaries, or simply war profiteers (often Iranian or paid by organiza-tions in Iran) infiltrated the Iraqi police force and gave Iraqis on the street money to place bombs. In one story told to me by several Iraqis in Amman, a militia group had stopped a bus and kidnapped all the Sunnis on it. Several days later, they stopped another bus and kidnapped all the Shia on it. I asked who the group was and they replied: "It was all business! Who knows who the group was? You and I and my nephew can make a group!"

Meanwhile, a practice of informing that was encouraged by Saddam Hussein's totalitarian regime morphed into a new iteration during the

2003–2011 Iraq War: a decentralized war market of informing and kidnapping came to thrive, providing streams of revenue for militias and gangs who co-opted various ideological apparatuses for their own uses.[28] One ethnographic interlocutor proposed that the antecedents of the war market of informing—selling one's neighbor's whereabouts for cash—might be traced to the 1990s: during the American sanctions of Iraq, a generation of youth in the slums, called the *'atāgah* (collectors, an Iraqi colloquial word derived from the word for antique, *'atīq*) stockpiled trash, particularly plastic, to sell on the streets of poor neighborhoods such as Sadr City. As economic desperation grew during the sanctions, the *'atāgah* began to steal and resell car parts. According to several Iraqis I interviewed, this impoverished class of collectors became particularly involved later on in the wartime practice of becoming informants, or engaging in the practice of *'alāsah* to the militias. The word *'alāsah* typically has a lower-class valence, as one of my interviewees pointed out: "Normally this person is from the lower class people looking out only for their selfish interests." Crucial class dynamics played a role here, and *'alāsah* might in part be understood as a kind of class warfare or class revenge—wherein the impoverished sell out wealthier neighbors to the militias and disrupt prior hierarchies. The practice was catalyzed by a range of motives—from money to sectarian, nationalist, or personal vengeance. However, according to my interviews, Iraqis who worked for the American military were disproportionately targeted.

The practice of *'alāsah* in its first meaning denotes mastication and eating; the word and practice first became institutionalized during the vendettas of de-Ba'athification and increasingly emerged on the Iraqi street around 2006 referring to the snitch, typically one's neighbor or acquaintance, who disclosed a person's location to the militias (to be kidnapped or killed), or *ate them*, for money. Within this logic, the selling of plastic evolved into the theft and reselling of first property and then the coordinates of peoples' lives. In the height of the most sectarian period in the Iraq War, *'alāsah* emerged as an often-used slang word, meaning "to eat."[29]

Ahmed, an interpreter who became a role-player, told me:

"It means to chew people up. It means in context to be an informant, to give intel [to the militias] about people who work with the Americans."

"And is that the only definition?"

"You can chew someone in any kind of way: I would chew you if I told your boyfriend you were having an affair. But real chewing in war is not just informing but killing: He has been chewed."

Hakim, an Iraqi in Amman who was kidnapped during the war after he experienced *'alāsah* first-hand, explained: "It's a practice of someone who is either part of or acquainted with the gang. His job is two-fold: first, to advise the gang on a good bite. The good bite is the victim. He is con- nected. The family and the victim know him." Ominously, the practice of *'alāsah* hinges on the idea of a covert intermediary (*'alis*) who has some kind of connection (acquaintance, neighbor, cousin) to the person who will be kidnapped. In order to "eat" someone and secure their location for the militia, you must learn their habits. Hakim said that this person would snitch to the militia and in some cases would even pose to the family as a concerned and previously uninvolved party who could serve as a mediator. Hakim explained this dynamic:

"The [practice of] *'alāsah* is very important to the gang, not only to the fam- ily. The gang is interested in money they can get from the family. In this case, they're not interested in killing, just in the money."

"And what is the role of the *'alis?*"

"The *'alis* [person] is the facilitator to keep the hope in the family that they could get their son back if they paid the money. So he would be the man who would introduce himself as the helper. But he is actually a collaborator with the gang."

It is perhaps the practice of the *'alāsah* that most reconfigured social rela- tionships during the Iraq War. Secrecy between neighbors and acquaint- ances colonized spaces of prior comparative comfort and ease. Indeed, in the case Hakim described, it is the *'alis*, the very person who has betrayed the family, who doubles to keep hope alive.

This practice of *'alāsah* particularly impacted Iraqi intermediaries, from multiple directions: they were both "eaten-to" the militias and accused of "eating-others-to" the American military. Although *'alāsah* generally desig- nates eating-someone-to the militia, the directionality of the act is poten- tially fluid: several Iraqis I interviewed claimed that Iraqis giving informa- tion to the American military also constitutes *'alāsah*. One evening in

Amman, I drank coffee with a group of former Ba'athists, some of whom used to work for Saddam's notorious *mukhābarāt,* his core intelligence agency. The men were in exile in Jordan, each having survived an act of *alāsah* and consequent brutal kidnapping and beatings. One gestured at his bright piano-key teeth, repairs from dental work. I asked the group to explain the meaning of *alāsah,* and they began by giving me the usual definition.

> One began: "The *alāsah* is when someone is a snitch."

> The other intervened: "A snitch to a militia or a party or, of course, to the Americans."

> One of the other men in the group interrupted: "No, you are not using the word carefully. It is specifically giving information to the militia, not to the Americans."

Although the fluidity of the term is contested (and this addition may have been for my benefit as an American audience), several other Iraqis interviewed concurred that the term could mean supplying the American military with information as well—a moment that revealed the ruptures in wartime translation itself. The potential gaps here constitute nothing short of fidelity or betrayal, survival or possible death by militia. In such moments of (mis)translation, speech itself might become "an uncontrollable instrument of war"—where the voice, how it is read and understood, and the possibility of a third party overhearing, makes translation devastating (Mojaddedi 2019, 505).

For Iraqi intermediaries, such an accusation converts them from cultural translators with aspirations for Iraq into informants and traitors. I asked Daoud, a former interpreter and current role-player who worked for the Americans for passionate ideological reasons, if an Iraqi interpreter can also be considered to have partaken in the *alāsah,* for giving information to the Americans. Daoud looked deeply wounded, explaining that those individuals are traitors, and he is not a traitor: "No! You don't understand! That is not the right meaning at all. The *'alis* is a traitor. And I would kill myself before I would betray my country." Although Daoud initially framed the question around loyalty and disloyalty to principles here, he then repivoted, explaining: "most of those people only did it for the money"—that is, they, unlike he, did not have true commitments. As

Daoud struggled to make sense of his own wartime choices and those of others, the ground was unstable. Deeply devoted to his own country, he saw his work as a translator as a passion project, an attempt to create meaning and communication amidst chaos. To different social actors, loyalty to Iraq means different acts, because "Iraq" itself is a contested space, co-opted by both Saddam Hussein and the Americans.

.   .   .   .   .

At the beginning of the war, Mohammad, the son of diplomats and a student in Baghdad, fled to his relatives' home in southern Iraq with few English-speakers: "The same day the troops got there, I went down and spoke with them and started translating." Mohammad remarked on the scant training the soldiers had received: "They had such a basic, like primitive understanding of Iraq. Like a high school student who just read a book a few years ago that he vaguely remembers." Into this void, Mohammad became an improvisational intermediary for the Americans.

He recounted the errant assumption of the American soldiers:

"[The Americans] did arrest operations at the time. And they came to the town and asked, you know, who is powerful in the town, and they said these two people are because they have these big houses."

"Wow," I replied. "That's quite an assumption."

"Their idea was that whoever was most powerful at that time under Saddam was a Ba'athist. It just tells you about how much they knew or didn't know, coming in."

Mohammad described how American soldiers seeking Ba'athists inadvertently arrested two influential and wealthy religious Shia figures in the community: "They went and arrested them. They thought this was part of Saddam's regime. And I was like, 'No, they are actually the opposition.' And they were like, 'No, but they have big houses.'"

Mohammad approached the soldiers to explain their mistake, a mistake that revealed the complexity of a politics the U.S. troops were ill-equipped to understand: not all the rich were Ba'athists and they had targeted the opposition. More complexly, indeed some Shia clerics including

Moqtada al-Sadr had worked with Ba'athists under Saddam and had only come out after his fall as opposition. As Mohammad negotiated with the troops, two thousand Iraqis from the community congregated outside the base to petition for the cleric's release, while threatening to use violence.

> "The [Shia population] was not going to take that. They already suffered enough under Saddam. The last thing they were expecting was for the Americans to come in and do just what Saddam even wouldn't do."

> "Yes, say more," I replied.

> "Saddam didn't arrest these people cause he knew the impact that would have; so even Saddam left these people alone and became friends with them or gave them a blind eye. So this was not a smart move at all."

Mohammad ended up mediating effectively, and walking out with the two clerics from the base, and there was a local celebration. After this event, the American military increasingly asked Mohammad for assistance and the Iraqi population of the village sought his mediation.

Similar to Mohammad, Salam, an interpreter who had been educated in the United Kingdom and returned to Iraq thereafter, had also worked to correct blunders and offer cultural access. When he interpreted for the Iraqi police force during an American training program, he explained:

> "On one occasion, the [American] instructor was talking about something about religion. He's saying I don't care who you believe in, if it's God or Buddha. But he put it in a way that's kind of offensive. So I told him, 'You need to understand that religion has a great value in people's lives in Iraq.' I told him, 'I don't think I should translate what you just said to the Iraqis. Let's rephrase it.'"

> "And how did he react to this intervention?" I asked.

> "He did appreciate it. I didn't say it as if I was showing him he made a mistake. I was on his side."

In these and other instances, the military vision of mobilizing a cultural intermediary to produce soldiers as insiders, and to even slightly lessen the weight of their presence is at least momentarily or partially enacted: American soldiers are furnished with knowledges on the ground, in some instances avoiding blunders or changing an outcome with the Iraqi population they are trying to court. However, meaning-making became increasingly fraught as Iraqi resistance to the occupation escalated and American military doctrine emphasized that adversaries were embedded among the population. Many of the American forces, following their cultural trainings, read Saddam's rule, subsequent sectarian strife, and insurgencies in the country through a script of primordial division—rather than "as events that have been entangled in national and imperial power dynamics" (Saleh 2021, 28). The wife of a former interpreter explained a collective Iraqi sentiment: "[With time, we became] very, very disappointed. To tell you the truth, we felt that we were . . . such naïve, stupid people, as if we handed all our dreams for the future over to [the Americans]."

A brief ethnographic excursus shows how the position of Iraqi intermediaries bears layers of complexity, as they increasingly became viewed by other Iraqis as U.S. military informants. One winter day in 2013, two years after the American military withdrawal from Iraq, I sat in an ornate Iraqi restaurant in Amman, with a fountain and gilded sofa benches. The owner, a gregarious and very wealthy Iraqi man, took me to the back to show off a fish-pool and the giant spherical clay ovens to make Iraqi-style bread. It was one of my last days in Amman on this particular field trip and most of the Iraqis I had interviewed lived in the poor neighborhoods, didn't have official residency permits, and could not work legally. During this first glimpse into the expatriate Iraqi elite, I was told by the owner of the restaurant to look across the room: the translator was here—it was "Dr. Hakim," who had worked closely with Iraqi translators under Saddam Hussein. Dr. Hakim was dignified and professorial looking. The owner of the restaurant walked me over to his table:

> "This is young Miss Nomi. She wanted very much to meet you. She is a student of Iraq." I told Dr. Hakim that I was writing about Iraqi translators and his eyes lit up.

> "American action has deprived Iraq of its translating community. I'm trying to establish new translation projects and to restart the process."

Dr. Hakim explained that he had worked in the translation division of the Foreign Office for the Iraqi Foreign Service:

"I was entrusted with improving the job of translation. Translation is a very important vocation for culture and public relations of a country. We created new generations of young translators. The problem was, they became informants."

It was at this moment when Dr. Hakim's lunch guests arrived, interrupting this turn in our conversation; he agreed to meet with me the following evening in a café. The next evening, Dr. Hakim described the translators of the Iraq War—many of whom were his former students or came from the young cohort of translators he had cultivated while working in the Foreign Office—as "victims of circumstance, who worked with the Americans because they needed a job."

"Is it possible to work as a translator for the U.S. military," I asked, "and not become an informant?"

"No, they couldn't, because the American soldiers wouldn't even respect them. The [soldiers] treated the [translators] like dirt. You've got to do this [give us information], and that's it. They treated them like dirt. That's what all my students told me."

"And are mediation and cultural translation ever possible in such circumstances? In war, do you always have to choose one side?"

"Cultural translation," he replied, "has its avenue, a beautiful avenue, and it's full of nice trees. And cultural translation is a civilized vocation. You can do it without being an informant."

"Were you yourself able to attain this ideal during the war?

"I practiced pure translation [during the war]—negotiation with the American forces represented by a group of people who elected me to represent them. I represented my case with respect."

However, for Dr. Hakim, such an ideal was, in general, very difficult to attain in wartime. He contrasted his own experience with what he observed about the experiences of some of the other Iraqis translators:

"What we are talking about is people betraying their society to a foreign power who may claim to have brought peace, but you can't be sure about the kind of peace you're talking about."

"And what are the consequences?" I asked.

"When you help them in this process, you're not actually doing cultural translation or interpreting; you're doing a service that may not be an honorable service. That is why those people, in the eye of the terrorist, in the eyes of the resistance, in the eyes of those who rejected them, were traitors and deserved to die."

According to Dr. Hakim, these Iraqis were sometimes rejected because they became invested in an offered peace without knowing its terms, invoking Sartre's writing on the collaborators with Vichy France. Suffering from "historicism," collaborators imagine a future where the occupiers will prevail, a metaphysics that conflates *is* with *ought,* and validates events "simply because they occurred": "It represented a subtle form of escapism. By jumping forward a few centuries and, from thence, looking back on the present to contemplate it from afar and resituate it in history, one helped turn it into something past and disguise its unbearable consequences" (Sartre [1945] 2008, 55).

Even though Iraqis who worked with the Americans did so for a complex range of reasons—and the question of loyalty was fraught with ambiguity because Iraq itself remained contested—as the condition of Iraq deteriorated, an Iraqi consensus developed. In that consensus, loyalty to Iraqi's future required resistance to U.S. presence. Dr. Hakim was passionate about translation as an ideal. He spoke with the most fervor as he imagined cultivating a new generation of literary and cultural translators to cross borders. He believed in an Iraq where this was possible. Meanwhile, he expressed the nuance and complexity of the impact of war on translation, and, more specifically, the choice of some Iraqis to work with the U.S. military.

A far less nuanced version of this Sartrean perception among the broader Iraqi population perhaps in part worked to convert Iraqi intermediaries into a new social category of collaborator. Regardless of their actual wartime motivations, Iraqis who worked with the Americans were increasingly seen as individuals who lacked commitments and might be "doubles"—working for both sides for gain alone. Indeed, as the war progressed, the critical gaze of both other Iraqis and of American soldiers projected duplicity upon Iraqi intermediaries, accusing them of being traitors and spies, and even more frequently of "only doing it for money," or potentially having "nothing" underneath—that is, of being radically unpredictable agents who would change sides on a dime. This particular convergence of gazes launched Iraqi intermediaries into a space of heightened performance, wherein they had to constantly act out and negate various affiliations and identities in order to survive.

As the 2003–2011 Iraq War progressed, the position of Iraqi intermediaries became increasingly fragile. Some other Iraqis resented their countrymen's new flow of income and others increasingly accused them of being traitors. Like many Iraqis, Um Umar, a middle-aged Iraqi woman in Amman, understood this choice of employment as a capacity to shift into any new role to further self-interests, driven only by the pursuit of economic gain, refusing to make a political commitment to their nation—that is, to inhabit a mask with nothing underneath:

> "I was so against it from the start. You are an occupied country! Why would you work with the occupiers? What, you say to the occupiers, Welcome, I will help you occupy my country more!"

> "And were you close with anyone who made this choice?" I asked.

> "My cousin went with his children to work in the Green Zone. He was the first person to sign up. He used to work for Saddam's *Mukhābarāt!* My cousin has no principles."

> "And what is your relationship with him now?"

> "I still don't speak with him. You must have principles in your life. You must have values. Some people don't understand. They want a nice car, a happy life. They will work for anyone. There aren't ideas behind their actions."

These accusatory gazes also sometimes came from within the community of Iraqi intermediaries, as they contested each other's motivations and sincerity. In some instances, these accusations took on more of a sectarian cast several years into the war, in particular as differences of sect became reified by rhetorics under American occupation and "sect" became a catch-all for other forms of difference.[30] Some Sunni and Shia translators I interviewed believed those of the other sect were simply trying to seek American visas, or to push their agendas in a future Iraqi polity, rather than doing the work sincerely. These accusations took on more heft as sectarianism increasingly became "a social fact" as sect and place were conflated during the global war on terror (Rubaii 2019, 126).[31]

Simultaneously, American soldiers, even as they increasingly recruited Iraqis to act as the "eyes and ears" of the street, described their deep suspicion of the Iraqis they employed, a continuous reminder of the limits of "cultural translation." In particular, a stereotypical and orientalist vision perpetuated by military cultural materials, of Iraqis as having a culture that was somehow intrinsically deceptive and focused on profit, created barriers. Most soldiers described how Iraqi intermediaries were interested only in the money and thus could not necessarily be trusted. One explained: "They were only in it for the money. They would go wherever the money was." Another elaborated: "In the beginning, we were giving them the best money on the block. Then money came in from Iran, so we didn't know what they were going to do." Another soldier stated: "We would send this guy to the store and he would short-change us and get his own stuff. You have to look out for yourself first. Ninety percent of them do it for money." Another agreed: "They'll do anything for money: they'll pay to get officer status or a diploma. They're all Ali Babas; they'll sell diesel fuel cut with water. There was an interpreter stealing ammo from us. You have to assume anyone wants to kill you. It is bounty."

Other soldiers, rather than assuming a commitment only to economic profit, accused Iraqis of looking out for sectarian agendas and maintaining their own biases. A major described his experience: "I didn't trust any of the fourteen [translators]. It's one of my pet peeves. I don't want any Nationals to work for us unless they are divested of their old ethnic biases and embrace American ethnocentrism."

Some soldiers privileged particular groups or sects of Iraqis as intrinsically more trustworthy. One colonel explained: "The Christian population embraced the Americans almost immediately when we moved in. They were mostly middle class. So we embraced that group early on. They were the only ones we could trust." Another offered: "The only ones we could trust were the Kurds. They were the only ones who were really loyal to the Americans." Another soldier countered that he didn't trust the Kurds and that his Kurdish interpreter was only working for his own interests:

"Our interpreter was all about going out and settling scores. I felt like we didn't keep tabs on him. He had a mob mentality."

"Can you elaborate?" I asked.

"He was having conversations with Iraqis that he wasn't fully translating. He was going rogue on us. But that was his motivation. All he wanted to do was to set himself back up in Kurdistan."

"Fred," a Green Beret who spent twenty years across the globe interacting with locals and interpreters, explained he always assumed local intermediaries had their own agendas, namely because they wanted to protect their families and the Americans would eventually leave. He reminisced in particular on his time in Vietnam:

> "[The interpreter] has a thousand reasons to mislead the Americans. The Americans will give him enough steaks for the whole month. But the Viet Cong will give him half a hot dog to not kill him and his family."

> "That sure puts him in a bad position," I offered.

> "He is going to take the half a hot dog," Fred replied. "But he might also take the steaks."

American soldiers consistently engaged in discourses about betrayal and potential betrayal by Iraqi intermediaries—even though it was extremely rare for an interpreter to "turn"—and sometimes even when interpreters had proven their trustworthiness. One morning, I was sitting in a diner in an American town with two soldiers, Stewart and Patrick. Over the clink of forks as we ate our eggs and bacon and mopped our plates with biscuits, they began to tell me about their former interpreters. Stewart offered: "They always lie. If you're in front of them and their shoes are red, they'll say they are white. I don't trust any of them." Stewart then shared how he had killed his interpreter:

> "He turned on us. He called his homies. Supposedly they had kidnapped his family. He walked us into an ambush."

> "Who was this interpreter?" I asked.

> "He was there when we got there, a CAT2, no classified access. He interpreted documents. He had been with us for four months. I don't trust any of them. I consider them all to be hostiles. They're the most dangerous of all, because they have access. They sleep in our compound."

> "And what was your impression of this man?"

"I thought he was just a professional. He stayed in the same compound."

"And then what happened?"

"When he turned I was a little surprised. It was elaborate for them. They were waiting for us."

"How did you know it was him who tipped them off?" I asked.

"Intel after the fact confirmed it. At the time, he picked up a gun and pointed it at us."

"We knew," Stewart interjected. "You can't hesitate. When you kill someone, it's not the person; it's the job. It's you or me."

Patrick offered a counter-story, of trust: "We had a terp for two rotations, five years. He was a local. We treated him like he was one of the team." Patrick explained that nonetheless, one should assume that an interpreter is untrustworthy until proven otherwise:

> "We vet him all the time. We give him info we already know to see if he'll lie. We'll ask—already knowing—what the prayer was on Friday if he said he went. And he will lie because he wasn't there."

"What do you do next?"

"Once you get the first lie, you do not confront them. You start tracking him."

Indeed, Patrick explained, even if he does trust an interpreter, he assumes that information will move through him and thus one must use this to strategic advantage: "We use him. We'll mislead others through him. I plant false stuff with people I *do* trust." I noted that interpreters actually "turning" was exceedingly rare, and asked how the interpreter protected him or herself in such a dangerous climate, knowing that interpreters often wore masks to protect their identities from the militias. Patrick countered: "If the interpreter is masking his appearance, how reliable is he going to be? He is acknowledging that he is scared. His interpreting will be tainted. He won't tell us everything."

These stories point to a fragile climate, where wartime intermediaries' lives are rendered particularly expendable. Ostensibly, Stewart's interpreter had been forced to tip off the militias because they had kidnapped his family, and he was trying to protect them; his family had been kidnapped in the first place because he had worked with the Americans. Rather than acknowledging these conditions, Patrick assumed from the outset that interpreters would lie, and found their impulses to mask (a survival mechanism in wartime) evidence of disingenuousness.

Amidst this climate and their tenuous location in it, Iraqi intermediaries inhabited a delicate balancing act, which demanded degrees of self-protective masking in certain circumstances. In front of Iraqis they did not know well, some described how they conversely performed anti-Americanism (or for many, drew on their own ambivalence about American occupation in their interactions) in order to not be snitched on to the militias. In front of their American employers, they portrayed their fidelity to the American project (making sure to wear it explicitly on their shirt sleeves, even if they felt it) as they watched other Iraqi intermediaries accused of being spies. Some of my interlocutors described the imperative of *acting* as a wartime intermediary: the necessity of varying one's route to work, of wearing the veil even if they might not have done so otherwise, of concealing their work and lying about it when directly asked. For example, because the American soldiers and companies worked longer hours, a plausible lie was working as a teacher or nurse.

The majority of Iraqis who worked with the Americans did not share the details of their new lives with their neighbors, acquaintances, families, or friends. They explained that they could not trust acquaintances and even some friends to keep this secret, and they did not want to frighten their families or put them at risk. They were desperately afraid for their families, while also working desperately to feed them. Salem offered: "I did everything possible so I would not be exposed. You change your route, one day you go by walking, the next day you take a cab, the next, you go by bus." Ahmed, another former interpreter added: "You have to change the names in your cell phone: you have to take out 'Mom' and 'Dad,' so they can't find your loved ones. In this time, no one can be who they are." Additionally, they were careful to perform normative forms of Iraqi nationalism in front of their acquaintances and neighbors. Another former interpreter, Salem described: "They would try to test you by making a bad comment about the Americans. And you would have to say, 'Yeah, those dogs.'" One former interpreter discussed the terrifying existential implications of this in-betweenness: "You are on forever moving sands: there is nothing to hold fast to."

Mohammad, a former interpreter who worked as a role-player thereafter, agreed with this notion of a life lived on moving sands: "It is like a play in wartime. The terror, though, is real." Although Mohammad is describing a broader wartime condition rather than the particular position of wartime intermediaries, those who worked with the Americans were especially required to occupy this mode of performance in order to survive. He first described how it is crucial to compose a mask: "You must be controlled and calm. You have to keep your cool." He elaborated on how he would interact with Iraqis in the streets, particularly when he had to go through Iraqi checkpoints:

"You must control everything on your face. I had to go through an al-Qa'ida checkpoint with my Green Zone badge in my back pocket."

"How frightening."

"And from my name, you can't tell if I am Sunni or Shia. And in fact, my father is Sunni, and my mother is Shia. They asked, which are you?"

Because Mohammad's ID card indicated that he was from a Shia town, he had to make up a backstory fast:

"They wanted to know, who are you, why are you here? And I said, because of [the Shia militias], those dogs, we had to leave."

"What happened?"

"It worked this time, invoking al-Qa'ida's enemy, they said, 'Okay brother, peace be with you.' And then I had to make up opposite lies in a Shia neighborhood."

Mohammad explained that in moment like this, "the victim must cheat the criminal" by feigning comfort and affiliation with them, and casually embed favorable information about yourself in your conversation:

"If you are at an al-Qa'ida checkpoint, you must act out that you are not afraid. You say to them, how are you guys doing? Can I bring you anything, water, anything, from my house?"

"Did you then bring them anything?"

"This just shows I live here in this neighborhood. Do not show anything on your face. Tell them you are going to bring them water. Then disappear."

For Mohammad, although he was practiced in the art of the wartime play, acting as necessary in front of other Iraqis, he learned the hard way that one must do so in front of more intimate circles. During the height of sectarian violence, a distant cousin of a different sect snitched on him to the militia.

This necessity of heightened performance also saturated Iraqi intermediaries' interactions with Americans. Firas, a former interpreter, explained: "I wanted to show them, I am with you brother. So I would try to talk about Hollywood movies. I would say 'Fuck' all the time because they did, too. I wanted them to trust me." Salem described a similar process: "I wanted to do whatever I could to show I was one of them. I wanted them to feel like we were all brothers." Salem had taken great pains to show American soldiers that he was trustworthy. He noted the pain and embarrassment he felt when an American soldier expelled him out of the cafeteria on the military base, even though his badge indicated he had access to lunch:

> "[The soldier] approached me and said: 'You have *no* right to be here. Get out of here.' I told him—I was very embarrassed—I told him, 'Ok, I'll just take this plate of salad and leave.' As I was putting the salad on my plate, he returned and said: 'What did I tell you?? You have no right to stay here.'"

"How awful."

> "And there were two Americans, two other Americans standing in the line, and he addressed them and said: 'Gentlemen, will you tell this guy he has no right to be here?' They saw I had a badge afterwards and apologized to me, but after that I didn't feel like having lunch."

Iraqis turned to work with the Americans for a range of reasons amidst desperate circumstances: seeking a better future for their own country, for their families, and for sustenance in wartime. In the dream, the ideal, of cultural translation and mediation, a neutral go-between conveys the concerns of both parties to each other. The intermediary is not compelled to take on a political position in relation to either side. In practice, on the ground in wartime, mediating-*between* is reconfigured as mediating-*for* one's employer to serve his particular politics and objectives. Ultimately, as Iraqi resentment intensified, the future "Iraq" Americans presented to them appeared increasingly dismal and traitorous: Iraqi infrastructure had been demolished and never rebuilt; after the American disbanding of the Iraqi army and security forces, security was flailing and legions of unemployed Iraqis were joining the resistance; the majority of the Iraqi population lacked electricity, clean water, and medical care; and Iraq's oil wealth was being plundered by American companies.

As more Iraqis turned against the American occupation, Iraqi intermediaries were converted, without their own consent, from hopeful mediators and translators to U.S. military technologies, gathering data about other Iraqis for the Americans. Namely, they were transformed from inhabiting ethical subject positions (seeking to improve their country) into reified instruments (co-opted for American use). As the political situation in Iraq worsened, Iraqi intermediaries indeed lived on *moving sands* —accused, deemed dispensable, and required to perform in order to survive. In the process, the U.S. Department of Defense's wartime ideal of a stable form of embodied knowledge offered by a cultural intermediary often disappeared beneath that same quicksand.

[Field Poem]: The Quadrant

Climb
in, climb out of the little black square—

The village rises into form between the pines.
Cows and goats stand in the forest. Flimsy wood
and storage containers: Muslim quarter   Christian
        quarter   Assemble/
disassemble. At a military technology fair in Orlando,
you can purchase a village in a box.
Just add people: live inside it for a time.

[Field Poem]: Driving Out of the Woods to the Motel

For your second job, you're a parking attendant or a poultry process
worker: stun and kill them, trim them and cut into portions, bone and
weigh and grade them. You are a hotel maid. If an American soldier
stays in the room you clean, you will fold his uniform as crisply as love,
a message that you too call it a *liberation*. Your brother calls it
an *occupation*, tells you: Do not become American. Brother,
the sanctions: 2 kilos sugar / 3 rice / 1 oil / 9 flour parsed into
sections? Buy lipstick at the drug store. Watch Ramadan soaps. Number
your hungers. Braise the bird until it is gold with lemon. Unstring your
wish:

one bone liberation,
one bone occupation.

3 The Theaters of War

There are animals in this wood. Goats mewl in the Muslim village. We cross the leafy footpath, past the tiny bodiless cemetery, into the Christian village. I am carrying a green apple in my pocket to feed Salma the pig. I am led by Laith, an Iraqi former interpreter who has skulls, laced with eagles, tattooed on his arms; and Jimmy, a sunburnt local wearing overalls who likes to go fishing in the nearby lake when he gets time off from role-playing. During the war, Laith had to wear long sleeves. You had to keep secret what you did with the Americans. Together we comb the woods for Salma. She is old, this pig, and they say she is a little sullen. Sometimes she climbs into a pocket of coolness among the pine needles, under the fake hospital with the sign (written in Arabic letters) *Mustashfā*, but today we cannot find her. We three share sunflower seeds, tossing husks into needles, calling the pig's name.

We are an unlikely meandering trio, carrying the lure of the apple among the simulacra—a military miniaturization of nation that feels like a twenty-first-century World's Fair.[1] The players in this game have variable relationships to country, self, and other. Laith risked death daily in the war, like many of the others here. Now he is playing a Muslim villager in a war game. Jimmy is a local role-player who has lived around these parts for a long time. Now he plays a Christian villager in a war game, encountering Iraqis for the first time in his life. Often in the various training camps, there are opposed villages, or a village divided in half—Muslim and Christian, Sunni and Shia. Meanwhile, complex (actual) lived histories fold leaflike in each chest, and are intermittently felt and made present. Power structures and tensions, the shifting political economy of the predeployment exercises, and the dark hold of the past sometimes encroach here, in a strange continuum where there seems to be little break between the simulated and the "real" (Boellstorff 2008).[2] But in other times, we are simply together, together only here in its flatness. And everything dissolves but the green enclosure of this other-world, as Baudrillard would have it, seemingly, uneasily referent-less ([1981] 1995).

Meanwhile, as I wield my books of theory and my notebooks—the scaffolding of quotations of scholars and philosophers to buttress my seeing— I know that I too am not exempt: I'm likewise a translator between lexicons, a wearer of a mask in a contested game. We each move between worlds variably, as I wonder how to best write about ethical actors (and myself be one) in these woods, the sky above us razored into triangles by the branches.

After we do not find Salma, we disperse. There are little hubs of activity in the woods. On one deck of a wooden house constructed for the simulations, some of the Iraqi men play dominoes and smoke. In one of the usual tensions that inflects the space, the domino-playing is both a break from work (an exit from the simulation) and an ongoing representation of hobbies pursued in the Middle East (where the real and the performance double back on each other). The women sit on another deck and talk and smoke. Or you can get your hair trimmed or your eyebrows done in a chair in the woods. Across the way, another role-player is deftly threading eyebrows and upper lips. I squirm and yelp as he leaves naked red patches on my face, making me, they say "much more beautiful," and laughing, "Nomi Basra, you're a scaredy-cat!"

There is still plenty of time, the milky light falling through the trees. Another role-player and I walk into the little sets, the flimsy rooms carefully ornamented with clues for the training soldiers: a poster of the leader of a militia; or a prayer rug covered in a little scattering of bullets (blanks);

the imam's chair like an elevated throne, a blackboard in the adjacent school drawn with two planes flaming into the side of a building. These props mean the imam is in cahoots with the militia, and it is the task of the soldiers to procure this information. The role-player I am with is in costume to play the imam. He is cloaked in a giant black robe, his head swathed in a white *kūffiyah*. He is still wearing his sunglasses and clownishly, he pulls his hoodie over his head. "Ha, you look like 50 Cent! Play his bodyguard, Nomi," they tease me. "Take the M16!" The emptied gun is very heavy across my lap. I sit at the base of the imam's chair. The role-players I am with are again acting out the militia that nearly killed them in the actual war. I brandish the gun. Click: again, we aim our iPhones at the scene. In the first, I pose. I play-act a girl with a gun, and they laugh. In the second, my face is colorless terror. We are always in between acts. We are never in between acts.

. . . . .

Each month, the role-players move between the mock villages and the "real" world, where they hold down other jobs, from construction labor or service industry work at restaurants or hotels, to teaching Arabic on the side. As in their prior work in Iraq for the U.S. military, these individuals were drawn to simulation jobs for a range of reasons: most, trying to get themselves established in a new country, were lured by contracts available to Arabic speakers and the possibility of a stable future for their family. As one role-player told me: "We get $250 a day to just sit around in the woods!"[3] Another concurred: "Of course, it's really good money. We're trying to get settled here." Some expressed a double patriotic impetus, hoping to serve their new country by assisting the U.S. Army with Middle Eastern knowledge, and thereby helping the Middle East. Further, as several role-players explained, doing this work was a means of counteracting American military perceptions of Arabs: "Every American who hasn't been over there thinks that every Iraqi is savage. They are getting know that Iraqis are normal people." Another role-player explained: "I want to give them a real picture of us. Like, we're not bad. We're not Bin Laden. I want to give them the picture that we are peaceful people, we love to love and be loved. I want them to feel we're human like them. We're not 'third world,' no."

Others expressed painful internal contradictions about their chosen work, or more frequently suggested that others experienced such contradictions, particularly in the early years of the simulations during the height of the Iraq War. One role-player described this predicament: "They were saying to each other, why are we helping these murderers against the Iraqi people?" A strongly pro-U.S. military Arab American who had worked to hire the role-players and had also himself role-played, described the role-players thus: "Half of them were prejudiced, anti-American. Love it or leave it. Why are you here, you Son of a Bitch?" Indeed, the role-players each negotiated different imaginaries of their own futures: some sought to anchor themselves in America, while others imagined returning to the Middle East after securing the privileges of U.S. citizenship.

The role-players were in a qualitatively new circumstance, now residents of the United States, en route to securing citizenship—thus effectively working for their *new* country, rather than working for a country occupying their own country. However, much like their prior work in Iraq, these individuals both inhabited complex motivations and felt the ongoing scrutiny of their countrymen. Ultimately, the "real" and the "act" interpenetrate in unpredictable ways.

As Iraqis actualize roles in the U.S. military's mock Middle Eastern villages and as they re-create themselves as (actual) new immigrants within the United States, they still inhabit some of the imperatives and the intensities of the 2003 Iraq War and its aftermath. Moreover, the spaces of the mock villages themselves augment some of the national and economic anxieties that these individuals experienced during the war. Logics from recent history and politics in Iraq (from totalitarianism, U.S. occupation, the previously noted "American sectarian approach" that fomented civil war, and wartime suspicion and targeting of U.S.-affiliated Iraqis) conjoin with new forms of scrutiny in the United States to create a wholly merited hermeneutics of suspicion. As the FBI increasingly recruits Middle Eastern immigrants in the United States to generate profiles of their countrymen, pitting them against each other, an impression of potential surveillance and a lack of trust sometimes infuses relationships between those who don't know each other well. Moreover, working within the mock villages sharpens this anxiety around potential doubleness of others, as role-players negotiate the theatricality of the spaces while they compete for material resources and social capital within them. As those who recently lived war repetitively act it out, certain after-effects of the referent—namely wartime anxieties around national affiliation—collide with economic uncertainty to complicate performances. Here, a caveat, as later pages will discuss, the archetypes that populate the mock village— from the representations of markets and mosques to the insurgent and mourning mother, indeed create a comedic gap from the referent (Iraq and the war). Yet although these *military representations* of the referent typically collapse in on themselves, there are certain *social modes*, carried over from war for the individuals acting out war that persist and occasionally encroach.

If the "performance of everyday life" (Goffman 1959) became even more amplified in Iraq during the war, as a means of survival, the choreographed performances of training sites in America presents other questions and stakes. Rather than inhabiting any kind of prescribed or secure sets of identities, these individuals often inhabit the demands of performance, both within and exterior to the explicit modules of the simulations. Fluid and complex inhabitations of identity and nation seep into *both* the formal enactments and the in-between times, in the mock villages after one act ends and the next one begins; and in the weeks in between the trainings.

In the simulation of war, that prior action—the referent—encroaches in unpredictable ways, undermining theories proposing that the simulation generates conditions that entirely abolish the thing-itself (Baudrillard [1981] 1995, Der Derian 2009).[4] Ongoing performance and renegotiation of nation and identity coexist uneasily with a military mandate to produce and ultimately contain identities in the trainings. In training simulations, military personnel often render complex and overlapping series of identities—that manifested changeably within different contexts—as violent fixities in order to extract information out of them. However, meanwhile, actual lived identities of Iraqi intermediaries, like those of all human beings, are continuously renegotiated.[5] Empire creates the frame here, conscripting individuals as human technology to perform its fantasies, without shielding them from the costs.

·   ·   ·   ·   ·

One afternoon I was sitting at a picnic table with an Iraqi role-player friend, and he told me that a few weeks prior he had been approached by an FBI agent who wanted to employ him. He told me that right away he had looked the man up on the internet: "I wanted to see if he was who he said he was. What if this was someone playing a trick, trying to test my loyalty to America?"

The job in question was to monitor the nearby Iraqi community in diaspora—or in some cases, Arabs more broadly—to track allegiances and politics, to see if there was anyone suspicious. My friend was reticent and turned the offer down, unswayed by the money that was offered. This conversation echoed many similar conversations of Iraqis approached by FBI agents in the United States, asked to spy on their countrymen either in the United States or in Iraq. Another friend who had been approached told me he had also turned them down, explaining to them, that after years of working as an interpreter during the war: "I'm tired now. I just want to live."

Another friend who was approached was asked to spend time in a Shia mosque assessing peoples' leanings and their possible links to Hizbullah, also turned down the work, explaining to the agent: "I'm sorry. I'm tired now." I knew that my friend wasn't interested in such work and did not want to spy on anyone. He joked to me: "Not only that but they would discover me immediately. I never went to the mosque! They would wonder what I was doing there." These young Iraqi men were in delicate positions: working towards citizenship in the United States, they worried that the job request was a test of loyalty, even if they weren't interested in doing the work. They each declined to do the work, not wanting to infiltrate their own communities in this way. In these instances, they each offered a model of refusal to be co-opted, a reassertion of their own humanity.

However, this structure of Empire had already seeded within the community: the militarization of each against the other, and everyone suffering the fallout. These tactics of division by the state, combined with the aftermath of war, and a context of precarity and economic insecurity, create a space of great anxiety and lack of trust for Iraqi immigrants. Under these fraught conditions, it was extremely difficult for Iraqi role-players to completely trust either Americans (some of whom were weaponizing them against each other) or other Iraqis, making resettlement challenging. One resettlement official I interviewed described the fear and divisiveness of the circumstances, explaining that resettlement agencies in the United States often hired non-Iraqis to assist the Iraqis, in order to not trigger their anxieties and suspicions.[6]

Thus, and crucially: it is simply not that Iraqis "do not trust each other"—the sort of statement one might expect from U.S. military training materials. Rather, sometimes Iraqis who do not know each other (and have not yet situated each other's political positions) hesitate to trust readily—in a climate where they have been militarized against each other by the American state. As I will further show, this climate of anxiety in some instances turns to complex performativity for Iraqi role-players amidst the violence of military identity that was imposed upon them.

·   ·   ·   ·   ·

One afternoon while eating *bamiya* (tangy okra stewed in a broth with pomegranate syrup and lemon, served with rice, and on some days, meat) with Iraqi friends who worked as role-players, I asked them if they were friends with other Iraqis in the United States. Daoud replied with a proverb:

> "Even if a person becomes gold, don't put him in your pocket" [*ḥattá law yasīr dhahib, lā takhīlh fī jībik*].
>
> "What do you mean by that?"
>
> "The person [the proverb describes] is bad. He will hurt you. Don't trust him. He will make a hole in your pocket. You shouldn't forget this."

From the outset, one must assume not only the artifice of that gold, but its perilous and punishing properties: it will burn a hole in your pocket. In the kitchen, Daoud's wife had been stuffing the grape-leaves into slender cigars on a silver tray, and she came out to join in the conversation.

> "Are there other proverbs about this?" I asked them.
>
> Daoud replied: "In your face is a mirror and the back of your head a splinter" [*bil-wajh mirīyah u bil-gūfa silāya*].

Daoud explained that the face a person initially presents can rarely be trusted, after what had transpired in the war and its aftermath. Many of those interviewed said they did not interact with other Iraqis outside of work. Indeed, as diasporic Iraqis I worked with try to anchor themselves in a new country and use all their resources to secure their futures, some viewed other Iraqis as potential competitors or saboteurs. While the Iraqis I interviewed all narrated their decision to work with the American military as driven by ideological commitment to help both countries, or a way to help their family materially, the specter of other people who were driven solely by economic gain and personal interests haunted them. These individuals were called *maslaḥaččin*. A role-player named Janan explained:

> "They are *maslaḥaččin*, even more here than they were over there. They have nothing here, so they have to use anyone they can to get ahead."

> "What do they do?"

> "They want to destroy someone. They do not want to see him happy or succeed in his life. They are not happy here. They have envy. They have hunger. You might get eaten. They never say, *al-ḥamdulillāh*, it's enough."

Janan was not the only of my interlocutors to use the idiom of hunger and eating to describe life situations, from living under Saddam Hussein's regime to the succession of wars that followed.[7] Although idioms around eating and consumption are common throughout the world, the regularity and the particular contexts of their use may provide a window into a structural climate of both suspicion and vulnerability before and during the Iraq War.

An Iraqi Christian role-player recollects the first Gulf War: "In the 1990s, the [Iraqi] Muslims said we [the Christians] were Bush's party. But all of us were eaten." That is, all Iraqis, without regard to sect or persuasion, were ruined. If Iraqis were eaten by the state and in some instances by their neighbors during the informant culture of the Saddam era, that hunger was particularly commodified during the practice of *'alāsah* ("eating" or informing on others for money, as described in chapter 2) during the 2003–2011 Iraq War. It is this period in which one ate others into the ingesting militia or into the ingesting army.

Some of the Iraqis I interviewed felt it was always possible for their livelihoods, their dreams, and their ambitions to be cannibalized, so fragile were their newly transplanted lives. The idiom of being-eaten was used to describe the threat of small acts of sabotage or undermining others in the community. My interlocutors described these acts as often being the product of an embittered envy (*hasad*), that was also accompanied by hypocrisy. Another Iraqi role-player explained: "They have envy. But in front of you they say good job. Behind you they talk and talk. I don't want to be near them. They try to ruin other people."

Meanwhile, as many Iraqis deplore the *maslaḥaččin*, those who pursue only their own interests and hence "eat" others in their path, there is simultaneously the cautious protection of one's own hearth and interests—especially amidst the fear that they could be readily capsized. In another proverb Daoud shared, I was told one should try to offer help to new refugees with a pure heart:

> "Do good, and throw it to the ocean." He explained: "Do it for its own sake. Don't accept it being given back to you."

Another Iraqi friend in the room immediately undercut the proverb, and one interjected:

> "Why help those Iraqis? They won't do anything for you. They are out for themselves."

> Daoud replied with another proverb: "If your friend is sweet, do not eat all of him" [*idhā ṣadīqik ḥalū, mā ta'akulhu kulhu*].

In true friendship, Daoud explained, one must not take advantage of the other by consuming him.[8]

·   ·   ·   ·   ·

This fear of being consumed by the other in wartime or its aftermath sometimes led to carefully curated and opaque presentations of the self. One Iraqi individual who had lived in Amman and is presently in the United States recollected: "In Jordan, when one person is accepted to the International Organization for Migration [for immigration or asylum to the United States] they do not say anything because they're afraid of envy. Someone will ruin it. We had to keep certain things to ourselves."

This impulse around concealment and performance as a form of self-protection also persisted for some in the United States. Moreover, the spaces of the role-plays in particular engendered carefully crafted presentations of national identity and affiliation, both as a realm of potential social scrutiny (among Iraqis) and as means to secure future work (with the American military). Indeed, in the small and flourishing political economy of the cultural role-players, there was continuous anxiety over turf and which Iraqis were getting more contracts per month. Many newly resettled Iraqis voiced resentment that other Iraqis did not tell them about the availability of role-playing jobs when they arrived in the United States. One role-player complained: "No one told me about this work when I first got here. They will all do anything to keep it and keep others from getting ahead of them. If I had gotten in on this work at the beginning, I would be chosen [for contracts] over and over."

Many Iraqi role-players felt there was an *in-group*, an elite group of Iraqis who were hoarding the work, protecting their friends, offering gifts to those with more power in the hierarchy, and deflecting those with less social capital. Those who felt they were outsiders described the ones in the interior as *maslaḥaččin*, willing to do anything, including trample the interests of others, in order to preserve their own interests. Individuals who believed they were on the perimeter also accused the insiders of possessing *wasṭah*, an unearned form of nepotistic influence. In this constellation, both Sunnis and Shias were alternately accused of having *wasṭah*. Frequently in discussions of which group possessed the most *wasṭah*, there were slippages between the small political economy of the cultural role-playing job and the larger political context of Iraq. In some cases, the sectarianism that had been produced as a social fact during the war on terror became salient (Rubaii 2019, 126). One role-player, with a Shia background, asserted:

> "They only hire Sunnis who are former interpreters. They have *wasṭah*. But, they are backwards. The war did it to them."

Another role-player, with a Sunni background, stated in a separate conversation:

> "The [role-players] are all Shia. The Americans think the Sunni are the problem, and they are wrong. The Shia make a different picture [of what

happened in Iraq]. They act oppressed, but they have *wasṭah*. Why did they even come here? They have the government now."

Meanwhile, the Christians felt convinced that the Muslims had *wasṭah*. One role-player noted that a Christian had not been hired back:

> "She role-played only once . . . I was the one who recommended her. They needed women. And they never hired her back. Why didn't they hire her back? They discriminated and hired a Muslim from out of state."

To work in predeployment exercises, Iraqi role-players must negotiate becoming-American while they daily perform being-Iraqi within military terms. They are doing so in a climate, which, at its worst, draws on a history of wartime mistrust, and weaponizes them against each other within an American surveillance state, amidst the new economic and social precarities of being immigrants.

·   ·   ·   ·   ·

In the simulations, role-players are generally given guidelines and scripts of varying specificity and instructed to never "bring religion or politics into the box" on their own.[9] This prohibition sharply contrasts with many role-players' passionate interest in discussing religion and politics, particularly as they pertain to current events in Iraq. As for the military requirement, although religion and politics may be interwoven into scenarios in mission-specific ways, role-players must not bring in *outside* forms of religion, politics, or affiliation into the space of their employment. Yet despite these prohibitions, both the performance of the scenarios themselves and the interludes between them offer an opportunity for role-players to subtly enact Americanness and/or Iraqiness, yet often not in the manners prescribed by their employers. Through these enactments, they become socially legible to each other and themselves as new immigrants with a complex past, while generating ways of knowing and being, via their affiliations.

Although the mock spaces are designed for training the soldiers, the role-players are also inculcated in both Americanness (the object to defend in the construct of the military training) and Middle Easternness

(the object of potential threat and/or courtship). Although the simulations are at least partly scripted, the general milieu still constitutes a complex "nationalist pedagogy" (Bhabha 1990), and role-players may perform
variably and comment on those performances in a range of ways.[10]
Simultaneously, in between the simulations, the role-players continuously
perform affiliation for their Iraqi compatriots (alternately and contextu-
ally enacting loyalties to their mother country and to their new country of
residence) as well as for American military personnel, as a widely understood way of advancing in their jobs. Meanwhile, the majority of the role-
players are negotiating both affiliations, as Iraqis with green cards, cur-
rently en route towards the benefits of American citizenship.

The interpellation of the role-players into national affiliation occurs
within an intensified enactment of nation: the mock Middle Eastern war
simulation. These simulations offer an exemplary version of the imperial
impulse in the "ordering up of the world itself as an endless exhibition"
(Mitchell 1989, 218): a twenty-first-century panoramic World's Fair in
constant animation, wherein every object within it is positioned as an
excavatable meaning about the wartime adversary.[11]

Yet, even when people's lives are most starkly and literally rendered as
stage parts, such as in the simulations, role-players actively create their
own processes or signification and possibility as they perform affiliation.
Deploying "strategic essentialism" (Spivak 1988), role-players variably
embody and perform *Iraqiness* and *Americanness* both within and outside the simulations in order to get by and secure a range of interests in a
vulnerable period.[12]

Many of the Iraqis I encountered in America (both role-players or otherwise) discussed their anxieties about national or religious affiliations
and how they or other Iraqis displayed these identifications to each other
and were read. These performances were frequently charted via shifted
outward presentations—typically clothing. The wearing of the hijab
bears layered, contextual, and historical meanings (see, e.g., Mernissi
1975; Abu Lughod 1986; MacLeod 1992; Nasrallah 1995; El Guindi
1999; Hoffman-Ladd 1987; Mahmood 2004; Almila and Inglis
2017; Shirazi 2018), and in some instances became a particularly
charged and moveable marker for immigrants mediating their degree of
assimilation.[13]

For example, one Iraqi woman who was not religious and did not wear the hijab, told me that she was looking for a community, and she wanted other Iraqi kids to get to know her kids. However, she said that in order for that to happen, she had to "be different, to pretend to be religious, to wear different clothes." Another unveiled woman lamented: "I took off my veil in my own time. I didn't judge others for taking theirs off earlier or others for not taking theirs off. It is a personal decision; why is there such judgment?" One woman, who did wear the hijab, accused Iraqi immigrants of "saying one thing when they want something else. Think of their clothes. They wear one style over there, and then as soon as they come here, they change." These choices were often explained in relation to relative degrees of "Iraqiness" or "Americanness"—with the latter marked by promiscuity and consumerism. One Iraqi man who had proffered judgments about how some Iraqi women now dressed explained: "Some people are more American than others. I don't want my wife to go near those people. They are too American. I will never be American in this way. They forget where they're from."

One afternoon, I watched an Iraqi friend, Abeer, put on and take off her veil a number of times, as the contexts shifted. We were headed to the mall to go shopping for a dress for a wedding. As she got ready to leave the apartment, she put on her veil, and gestured at her pious choice this month, "Well, it is Ramadan." When we got in my car, we were discussing how it was to be perceived as Arab in the United States these days. She told me that she was always veiled in some neighborhoods, holiday or not, because she didn't want to be seen as less Iraqi than others. However, she had started working for an American company several months prior and she was the only Arab there. She was passionately determined to be the best employee she could be, and to be accepted, so she did not wear her veil to work, explaining of the other employees: "They don't all know where I'm from. Some of them think I'm Mexican! They don't know anything about Iraq." We entered the mall, and walked through its aisles, passing Old Navy and JC Penny, Victoria's Secret. The halls smelled sugary and floral with perfume samples and groups of American teenage girls walking by, giggling. Abeer swooshed off her veil in one movement: "I don't need this here!"

Identifying in distinctly "Iraqi" ways was also alternately praised and blamed. Some people referred to others as "*Iraqi* Iraqis" to indicate an authentic core. One woman, who considered herself more assimilated and cosmopolitan, referred to the same such people in disparaging terms: "There are some Iraqis here who don't want to become American: they have another mentality. They don't want to adjust here. They're not becoming American. They are completely enclosed in on themselves." It was explained to me further that Iraqis might alter their outward identifications depending on their interlocutor. For example, one Iraqi man told me of another: "He pretends to be a liberal in front of you, Nomi, so he appears more American; and then he changes when he is around other Iraqis."

These anxieties about affiliation played out in a range of everyday settings: on shopping trips, at school, and during a medical exam for the military, when role-players were upset that an Arabic-speaking nurse may have judged them for working with the Army: "It was none of her business that I do this work! I don't want Arabs knowing my business." This kind of anxiety around scrutiny and judgment did not always take on an explicitly national cast, but it was common enough. For example, in one Iraqi family, whenever there were interpersonal problems, members of the family often accused each other of having changed upon coming to America.

These anxieties appeared to inflect both the inner lives and outward interactions of many of these U.S. military–affiliated Iraqi immigrants, as they negotiated creating a new life in the United States amidst their wartime choices, the gazes of others (both Iraqis and the American state), and their desires to anchor themselves in their new country. Indeed, both the longer breaks between the simulations and the smaller interstices when role-players practiced and prepared right before the role-plays offered telling windows into these performances of affiliation and the anxiety that attended them.

During these interludes, when the soldiers are typically engaged in a mission in another portion of the village, the role-players typically engage in the following, still in the simulated village: smoking cigarettes or hookah, drinking tea, playing music, playing dominoes, occasionally giving haircuts, and chatting. In these settings, the role-players typically assume gendered formations: the men usually play dominoes on the porch, while the women sit and chat on another porch. Although the interludes are work breaks, the role-players remain within the fake village that is meant to approximate the Middle East.

Different communities contested the meaning of these interludes. In each case, however, what was at stake was frequently *Iraqiness* or *Americanness*. For some of the soldiers, the Iraqis hanging out adjacent to the mock houses were intrinsically an atmospheric addition to the set. One training soldier remarked on these interludes, describing the role-players as simply *being-Iraqi*:

> "They're good dudes. They're on their own timeline. We got here, and we're like, we have five days. And they were like not doing shit, just drinking tea."

> "And what are they supposed to be doing?" I asked.

> "I think the key [role-players] know how this will play out. The other guys [with smaller roles] are just being Iraqis, and that's awesome! That's what they need to do!"

On the contrary, the contractor in charge of the role-players did not necessarily perceive the dominoes as atmospheric flourish; according to a role-player, one of the supervisors cautioned them: "The boss says that if the Army is ever watching you, you have to stop playing dominoes; you better get to work right away. You better show them you want to be here."

A number of the role-players also perceived the dominoes as a meaning-laden activity, which potentially indicated level of commitment to the work and even, in some cases, relationship to nation. "Daoud" in particular was distressed by the dominoes. Daoud worked with the U.S. Army for passionate ideological reasons, and he sought to play his role with great commitment. To this end, he regularly researched his roles and sought to integrate as much finely tuned cultural knowledge as possible. He noted that some of the other role-players only wanted to play dominoes and hence resented his thorough approach. He explained:

> "Some of the other role-players got mad . . . because I was doing the scenario the best I could think of, but this made it take longer. They wanted to know why I was making it take so long."

> "And what did they want?" I asked.

> "All they ever wanted to do was play dominoes and take naps and not do work. They get really mad . . . when I make the scenario last any longer, they're like, come on, we want to get back to our game of dominoes."

"Salem," another role-player, explained an aversion among some of the Iraqis to Daoud's work ethic. Salem explained:

> "The reason they don't like this is that he takes the work very seriously. They tell him: 'Make it short! We want to play dominoes.' And he does not make it short."

For Daoud, avoiding dominoes and offering a proper rendition of his role was akin to expressing his support for the United States, his new country. He elaborated:

> "I believe in this work, but most people who do this work do not believe in it. Their hearts are not in it. Most people do this work just for the money, not for country."

According to Daoud, some of the other role-players are *maslaḥaččin*, motivated singularly by economic incentives. He interpreted this capitalistic mode, in part, as an insufficient ideological commitment to the *content* of the work. He and others of this opinion suggested that some of these *maslaḥaččin* were only securing American citizenship in order to move back to the Middle East to get a better job there. For some of the role-players, this attitude represented a refusal to commit to Americanness and their new country, while continuing to privilege Iraqiness. He and several others lamented the ambivalence of this other cohort of Iraqis in the United States:

> "Some of the Iraqis come here even if they don't like the American people. They eat and drink American food. The Americans help them. They say: 'We love Iraq.' While even still they live here!"

When I asked role-players if it was possible, rather, to have authentic and deeply felt attachments and affiliations with two countries, sometimes they expressed doubts. For members of a wartime population that had sought to mediate between Iraq and America—and had been read and often killed as American-allies—multiple allegiances had been severely punished. "Um Umar," whose cousin had worked with Saddam's secret police and subsequently with the Americans, explained: "You must choose. You can only have one angle in the world. You cannot be double-faced."

Amidst these tensions, the simulacral villages and the war games within them have the ongoing potential to become something like petri dishes—small, enclosed, intensified spaces for Iraqis to enact and rebuke affiliations. In his theories of play and fantasy, Gregory Bateson asserts that: games "place a frame" around a set of events, determining what kind of "sense" will be bestowed inside that frame" (Bateson 1961, 20). The creation of sense, however, compels mutual agreement, and play can only occur if the participants in the game exchange signals that indicate *play*. He offers the example of a pair of playing monkeys, using signs that resemble, but are not in fact, those of combat. That is, "the playful nip denotes the bite, but it does not denote what would be denoted by the bite" (Bateson 1972, 180). In this form of play, wherein the frame is clearly delineated, and the meta-communication is unmistakable, the nip denotes the bite and reminds the animals that bites exist, that they have been bitten, and that they might be bitten again. For Iraqis who have recently emerged from war and are now role-playing war within the charged national spaces of the simulations, "play" is likewise perilous and ambivalent. Bateson invokes as an example A. R. Radcliffe-Brown's work on the Andaman Island, wherein ritual blows of peacekeeping might be misread as actual combat, and the peaceful ceremony turns into bloodshed. He explains that such an instance leads to a far more ambiguous form of play, "constructed not upon the premise, 'This is play,' but rather around the question, 'Is this play?'" (Bateson 1972, 182).

In an essay which extends Bateson's, Goffman suggests that the frame that separates the wider world and the smaller social construct (of, in this case, the game) is "more like a screen than a solid wall" and can thereby alter that which passes through it (1961, 33).[14] For Goffman, the game's rules demarcate the very boundaries of reality, the rules that dictate what is relevant and what is real. Likewise, in American military simulations, only a porous membrane encloses the war games—and their surrounding activities and the subjective experiences of the role-players—from the war's afterbite. As former interpreters accused of collaboration, Iraqi role-players are still in an ongoing process of performing and concealing national sentiments and affiliation.

In the mock villages there are no stable "rules of irrelevance," *outside* properties of the game pieces that do not bear upon the outcome of the game (both the war game, and its interstices).[15] Theoretically, the outside lives of the role-players should be eliminated from the idealized space of the village. Yet, inside the mock villages, outside properties—in particular, anxieties about affiliations, sharp economic insecurities, and the surveillance enacted upon the community—continuously encroach. Any barrier to the wider world is fraught with leaks, and through it seep many of the present lived experiences of the wartime intermediary: the complexities of living as a postwar immigrant or asylee in the United States; the heightened anxieties about national affiliation produced by the spaces of the simulations; malaise around an American surveillance system trying to recruit them against each other; malaise about mistrust of other Iraqis and their judgments around assimilation, in particular through the lens of gender; meta-performances of affiliation in order to secure a future in North America; and essentially, war's bite, and its exilic aftermath.

.   .   .   .   .

"Najla" left Iraq after her family, some of whom had worked with the American military, was targeted during the 2003–2011 Iraq War. She had several traumatic memories from the war from when she was a child, but she was mostly focused on her future. She had lived in the United States for several years, working on her English and doing a range of tutoring and translating jobs online, but had only recently begun working as a role-player for the military. Najla was bright, independent, ambitious, and eager to begin a new life in the United States. One afternoon while we drank tea, I asked Najla about the atmosphere she encountered in the mock villages during a recent training.

> "[The training] was three days. It felt like a month. It was a nightmare. We had to cry and scream. We had the night shift."
>
> "And where were you in the mock village?"
>
> "We were in this one place that seems haunted. There were no windows. The ghosts talk with you. And I was working at night. There was no light. They gave us the scenario. It was severe."

For Najla, the training went awry in particular because she felt that another role-player, a man named Ekram, may have been trying to undermine her. Najla explained:

"He has done this work a long time, and we are new to it. He made it harder on us. They called on us to work there as a team. Why does one person come and ruin everything?"

She explained that in the past, she had worked with more seasoned role-players who had been willing to help the novices and give them guidance:

"With 'Alaa' [a long-time role-player], if anything you do is wrong, she tells you. Sometimes in the role-play she will tell us: 'Don't be silent, talk, talk, right now!'"

"How was that for you?"

"That's lovely. But this time [with Ekram], it didn't turn out this way. He wanted to make us do what he wanted."

Najla explains that she felt that unlike with Alaa, Ekram had not only *not assisted* the newcomers; he had made their time there more difficult, using them to advance his own professional interests:

"[The supervisors] gave us our names [for the scenario]. I forgot my name. There was no paper for us. [Ekram] just explained [the scenario]. He was the one with the paper. He explained fast, not clear. I said, 'Okay, let's go.'"

"Then what happened?"

"Okay, so I forgot the name. What did he do? He went to talk to the boss—he said [some of those people] don't know what to do. And I was like, what's your intention, to go to the boss and say something like that!"

In inhabiting the intensities of the simulacra, which evoked her (actual) emotions, Najla was rendered particularly susceptible to the ensuing unnerving braiding of the "real" and the simulacral as her affiliations were questioned. In this training as in others, over the course of the three days, the role-players had to enact the scenario approximately two dozen times. Najla thus soon memorized both her name and the arc of the scenario. She explained the scenario, which she described as emotionally and even physically taxing.

> "In the role-play, my [fake] Dad worked with thieves. I had a computer chip with secret information on it. In the story, my father gave it to me. My mother [in the role-play] was old and sick, and she was hiding a gun."

She described the entry of the (female) soldiers and how they aggressively searched the role-players: "The soldiers came and searched us. They searched us so much. It was rough. Some of them were very rough." I asked her if the training soldiers located the computer chip and weapons, and Najla explained the varying outcomes with the different groups of soldiers:

> "Three groups of them didn't find the chip. Some of them didn't find the gun either. So that was considered a bad thing for them, that they didn't find it."

> "How did the soldiers react to you?"

> "Some of them felt for us—they said, I know this is not comfortable, but I have to do this. Some of them, no! As soon as they come in, they say we come here to save the village. We know someone has something—like bombs. They showed us a picture of this bad guy. I said we don't know. We just stay at home. I said my mother is sick."

Najla told me that as the training soldiers stepped up their searches, she too began to react more intensely and become genuinely upset: "When they got harsh with me, I got harsher." I asked her, "For real?" And she replied: "Yes, for real. I had some real feelings. I was upset and annoyed. I felt it in my blood. It was harsh." Najla elaborated:

> "When they grabbed me, they wanted us to go [with them to be searched] one by one. I screamed about how I needed to be with my mother. I screamed, 'I can't!' To see what they would do."

> "What did they do?"

> "They grabbed me *so* hard with their hands. So rough. This happened so many times. I thought maybe this was part of the role. But I didn't imagine that the role would impact my body. I was veiled, and I said to not make me ashamed. To not lift my veil. But they did it anyway. They saw the gun, and they became more harsh with us. The act was good and not good: because *I lived it.*"

Amidst the intensity of this performance—which was both "good and not good," because she was *living-in-the-role,* a concept to which I later return—Najla was particularly startled by the lack of solidarity evinced by a more senior Iraqi role-player and his refusal to help her understand the ins and outs of the scenario.

Over the course of the interludes between the scenarios, tensions related more specifically to identity and affiliation manifested between the role-players. In one instance, Ekram also incited Najla about religion—creating a tension between inside religion (that which was enacted in the simulation) and outside religion (that which Najla might enact in "real" life)—thereby injecting into the space that which the rules of irrelevance prohibit. Although Najla did not wear a veil in her usual life, in her present role she was veiled. She explained:

> "The first day, I put on the hijab and I was veiled. And [Ekram] said something about religion. When the soldiers came to search me, I said, 'Do not touch my hijab: that would be shameful.' And after the role, Ekram laughed meanly and said, 'They don't even know it is normal for you to not wear the hijab!'"

> "How did you react?"

> "I didn't say anything, but I was like, what is his intention? And I felt he intended something about [my practice of] religion. But why did he bring religion into it?"

The issue escalated for Najla as the searches in the role-plays became more aggressive and Ekram reacted with mockery:

> "Before the end of the first day, one soldier annoyed me a lot. That soldier, her hands were going everywhere. Really! So many times. And when [the other role-players] told [Ekram] that, he laughed!"

> "How did you respond?"

> "I was so stressed. And I said, 'Why are you laughing'? This thing really upset me. *What,* I don't wear a veil [in normal life], so it's normal for them to see my body?! But when the soldiers were bothering [an actual veiled woman], [Ekram] got upset."

For Najla, Ekram was judging her choice to live as a more assimilated Iraqi woman who was trying to make a fresh start in America. Acutely aware of other Iraqis' shifting choices around self-presentation and behavior, many individuals read these choices in relation to degrees of abandonment or assimilation: that is distance from Iraqiness and approaching more Americanness. Najla's emotional response was provoked by an established trope of critique within the Iraqi diasporic community, such that a (theatrical) role became a catalyst for the ongoing negotiation of a shifting gender—as well as national—identity, all of which encroached within the space of the mock village.

Soon thereafter, in a rapid-fire series of events, Ekram had told the employers that Najla wasn't doing her job properly, imperiling her chances with the company. Najla shared the situation with other role-players, who initially commiserated and suggested complaining to the boss, and then insisted on staying silent. Najla was subsequently under the impression that other role-players were talking behind her back, leading to a mediation session with their employer. Their boss appeared to largely dismiss the situation with a warning, telling them that he was "an easy-going boss, and as long as you guys are happy, we are happy." Through the course of this drama, the spaces of the simulation were hardly pure containers for conducting a fixed role; rather, they were entirely porous, flooded with capitalistic competition, economic insecurity, complex performances of affiliation, opacity, and accusation—intensities of the outside that were integral to the role-players' wartime and postwartime lives in a new country.

.   .   .   .   .

While the military bases the "game" on the "real," certain forms of the lived, embodied, affective, and political real are deemed superfluous or divisive, exterior to the tactical goals of the game. This delicate tension creates a porous vertigo: Which forms and degrees of knowledge and emotion will be granted realness? At which moments are we inhabiting game-real? At which moments are we inhabiting the lived-real? When do they contort into each other?

Recent thinkers of the virtual have on the whole proposed that the image supplants the object. Jean Baudrillard insists: "The map precedes the territory" (1981, 11).[16] Paul Virilio (1989) cautions that target areas have increasingly come to resemble cinema locations.[17] In *Virtuous War*, the only prior ethnographic monograph that describes the mock Middle Eastern villages, James Der Derian examines contemporary war through the lens of Baudrillard, where the image precedes and eclipses the object entirely, and of Virilio, where there is a wavering back and forth between signs of the real and the real itself.[18] Der Derian describes his own experience of such indeterminacy while witnessing the experimental war game Desert Hammer VI at the National Training Center at Fort Irwin, California. "An M-22 simulator round, about the size of a fat shotgun shell, exploded nearby, as a Stinger missile crew fired at an F-16. Then came the white plumes of 'Hoffmans,' blanks that simulate the flash and bang of tank and artillery fire" (2009, 7). In the middle of watching this simulation, Der Derian and the other media got a radio order to move locations. When he instead paused to take a photograph, he found himself in the midst of the battle: "I froze, feeling the terrifying yet seductive rush that comes when the usual boundaries, between past and present, war and game, spectator and participant break down . . . I could imagine yet deny death" (8). Der Derian ventures the risks for soldiers who are trained in simulations: "Was there a paradox operating here? The closer the war game was able to technically reproduce the reality of war, the greater the dangers that might arise from confusing one for the other?" (14).

Der Derian situates all participants in the simulation within a uniform postmodern idiom, rather than assessing their potentially variegated relationships to the image and the object. In the course of his study, Der Derian interviews a range of experts, from Vice Admiral Arthur Cebrowski, who asserted that there was a contemporary revolution in military affairs (RMA) "unlike any seen since the Napoleanic age," to military officials onsite at Fort Irwin (2009, 130). In this latter case, he explains: "I had asked the stock questions: Would the friction of war overheat a cybernetic battle-plan? Would the surge of information overload all these digitized systems . . .? And I had received for the most part by the book responses: perhaps, but not so far, and besides, this is all in the experimental stage" (13). However, through most of his discussion of Fort Irwin, he relies on sound bites from military brochures: "digitization will get us inside the enemy's decision-making cycle" (5). To this end, we are typically left in a self-confirming unpeopled, disembodied, and mechanized environment where "each soldier carried a 3.5-inch computer disk in their breast pocket, not to stop a bullet but to store a digitized image of their predestined wound" (12). Most crucially, perhaps, Der Derian does not reprise any conversations with training soldiers and only one with an Iraqi mock villager. That individual offers only a brochure-worthy explanation of his motivations for doing the work: "We give the idea how does life look like over in Iraq, so we are doing the same thing here. To get the Marines trained good so when they go back in Iraq, we make sure they know what they are doing [and] how to deal with the people" (293).

These postmodern lenses hardly account for the complexity that is experienced by role-players in mock Middle Eastern villagers, who are intimately familiar with the referents of Iraq and war in the Middle East. Indeed, their current lives document the ways in which the vulnerability, suspicion, and accusation of war—as well as the after-effects—persist. These intensities are part of these individuals' lives, both beyond and, in some cases, impinging around and within the simulations. Moreover, not only have these individuals already been *there* (in the Middle East, the referent) for most of their lives, but after a foray into the image via their current employment, many imagine returning to the actual Middle East one day. That object (their home) may be altered via violence and shifting politics, but it is hardly obliterated or replaced by the military archetypes they themselves never believed were filled with any real content. That is, even in future encounters with the object, the image hardly acts as a supplanting force, at least for the role-players.

The experiences of the role-players within the simulations are perhaps closer to something like "syncopated time," where the past and present interpenetrate in complex ways (Schneider 2011).[19] Yet these interpenetrations of time amidst the reenactment of a historical trauma are not in this instance redemptive heterotopias offering the potential for historical redress and even reconciliation (Auslander 2010).[20] The simulations hardly disrupt or reconfigure preexisting power relations, as I will explore in subsequent chapters. Still, the simulations catch uneasy and potent intensities that continuously ripple into and borrow from ongoing life.

Najla is jolted in the simulation when she is searched by the training soldiers, yet perhaps even more so, when another Iraqi role-player needles her—an uncomfortable prodding about what kind of Iraqi woman she was and is, and what kind of American woman is she becoming, and what that might mean. For Iraqis with affiliations to the U.S. military during the war who bore the accompanying costs, these sensations are a particularly disturbing jolt. In such moments, aspects of the referent encroach and interface with the simulation. For many of these individuals, who are living in an America to which they, at grave cost, aligned, the past that punctures through is already deeply and inextricably part of the now.

Still, perhaps the near or total death of the referent holds in the case of a different military training simulation: 3-D immersive virtual reality trainings. Offering a panoply of options, from the simulation of flight to medic training to battlefield combat exercises, virtual cultural trainings have also been widely used by the military. Likewise, in these virtual trainings, soldiers are asked to question and engage in cultural interaction with Middle Easterners—in this case, digital avatars. Yet, in the training program "Tactical Iraqi," the avatars are "incapable of speech acts that are not scripted by the U.S. military"—thus, they cannot ask difficult questions about American foreign policy or the military (Gregory 2008, 18). A filmmaker I interviewed told me about visiting a military control room that incorporated digital technology of this ilk.

"We saw the control room. It was very intense. It was this room where this little guy was sitting on this roll-y chair with like sixteen screens in front of him, and he can control what pops up on the screens, because this is a facility that uses live half digital screens with avatars."

"And what did he make pop up? Did you see him do it?"

"He was like . . . let's pop up a woman and see what they do with that. Let's pop up a suicide bomber holding a kid and see how they respond to that. You know, this wizard behind a curtain, choosing what scenarios."

One afternoon, I drove with "George," a military contractor, down the dirt roads that lead to the mock villages. Part of the boom of post 9-11 security contracts, his company both facilitated exercises in mock villages and had a think tank in Washington, DC. He described constructing a training site: "It's a Field of Dreams," he told me. "We build it and hope they [the military] come." The company he worked for was also involved in virtual training courses that reserve officers take before coming on-site to the training courses. George told me that in some instances, (actual) Iraqi role-players are photographed and repurposed and turned into digital characters for virtual reality trainings. These digital icons are airtight human technology, acting out military imperatives, within an immaculate ventriloquism. Images inside an image, they move inside the frame, comply with all rules of relevance, and speak as coded. They are a foil to Najla and her knowledge and emotion—both about the country she left under duress some years ago and about her life in a new country. The coming chapters delve into how military personnel interface with and seek to corral and control that knowledge and emotion.

[Field Poem]: Creation Myth

(How the role-players came to speak)

Soldiers build our legs and arms:
newly made, we clap the fire
leg lift, leg lift, wake the ache
in the new hewn chest, circle
the pit, steep the tea, slice
the Spam into cheeks, break
the bread, then cry! Mouths
light open and shut
open, then shut, around
their words, their words,
their words,
words, words . . .

[Field Poem]: Driving Out of the Woods to the Motel

After the soldiers finish the game, neutralizing whomever they believe
is a danger to the free world, my friend & I drive out together, off
the highway, past a sign that says KIA. I say: "Killed-In-Action"? No.
It is a dealership, bright cars in a wide lot. As the city comes out
of the gasoline haze: Days Inn, Walmart, Chick fil-A (the woods bluing
to a point), we practice for his naturalization exam. Who is the "Father of
Our Country"? What are the principles of American democracy?
Renounce now, on oath, all prior loyalties. It is natural, friend, to want
to live. How neutral you wished to be, hired to bring your country to
life. No preparing for how the bomb packed with ball bearings & nails
denatured the body. The acronym, neutered, turns blank into a lot,
but how

we counted them,
row by row.

4 Left and Right Limits

We are in the woods, and it is almost midnight, as the training American soldiers rotate between a quadrant of rooms in the war game. In one room, a woman role-playing a mother weeps inconsolably and beats her chest over a tiny bodiless coffin: in the scenario, an American tank inadvertently crushed her child. In the adjacent room, when American soldiers visit the local police station, a woman who has been imprisoned for alleged affiliations with the militia cries out from the jail; she is then brutally beaten by the jailer. Military contractors running the simulation signal that she should cry out and then be beaten when they flicker the light off and on. In the next room over, the American soldiers visit the imam, trying to discover if he has ties to the militia. He is infuriated by the accusations and slams his hand on the table. In the last room, the soldiers meet with a governing figure in the village, discussing sectarian strife. The four rooms are decorated with orientalist props: tea sets, lamps, cloth, pillows, prayer rugs, and sometimes posters of militia members.

The quadrant's soundscape commingles: a square of murmurs, then voices rising; a square of grief; a square of beating; the hubbub of Arabic dialect again, seeming to act more as a confusing clamor rather than meaning for the soldiers. In each small theater, Iraqi role-players, hired as human technologies, become lamenter and imam; they become villager and mayor. In contexts one, two, three, and four, they are summoned to enact these continuums of living and dying, *as Arab* Others, an "Apparatus" to help soldiers realize their operational goals. Here in this enclosure of the pines, if the military personnel believe that soldiers have engaged incorrectly with the wartime locals, the game pauses; they reset.

In their essay on the culture industry, Theodor Adorno and Max Horkheimer caution that "hit songs, stars, and soap operas conform to types recurring cyclically as rigid invariants . . . the details become interchangeable" (2002, 98). In a lacerating enactment of the culture industry, human beings, too, are turned into interchangeable cultured-bodies—in this instance, circulating through a war market (and its exercises and games) where "the local Other" can be coded, known, and contained.

Meanwhile, an ironic disjuncture between military prescriptions for authenticity and role-players' experiences of inauthenticity generates moments of charged incongruency for those hired to embody constricted versions of their cultures. A charged tension manifests itself in the training apparatus: on the epistemological level, even as they experience excess, role-players work to make the simulations "look good" to retain their jobs. As they are asked to become cultural products and infinitely replicable archetypes, these wartime intermediaries enact complex epistemological labors, but not on their own terms.[1]

.   .   .   .   .

The contemporary aisles of the military's culture industry might have been Adorno and Horkheimer's nightmare, and the apotheosis of their concept: the culture of the adversary turns into a commoditized replica on a war market, a glossy product used to hone techniques of killing, according to a preordered rubric. One company that provides "cultural training and atmospheric training aids" to the U.S. military lists its culture products on its website, offering a chilling window into the culture industry of war. On offer are Mobile Cart Vendors (Fruit Vendor; Cheese Vendor; Butcher Vendor); Streetscaping (Monuments; Billboards; Cultural music; Village huts and tents); Graveyards ("5 Headstones assortment of sizes w/ Arabic Letters/Names"); Animal Carcass ("Nothing can be more realistic and stressful to our war fighters than seeing an animal carcass: Goat Carcass; Sheep Carcass; Horse Carcass; Chicken Carcass; Camel Carcass; Pig Carcass); and Human Corpse ("Human Body Corpse can be used for medical training or even cultural training exercises: Middle Eastern Corpse; African Corpse; South Asian Corpse; Asian Corpse").[2]

The culture industry is a "rotating" machine that amputates that which does not fit within preconstrued schema (Adorno and Horkheimer 2002, 106), and indeed, the American military can purchase a "Village in a Box."[3] In that box, there is everything from the just-mentioned atmospherics to the synthetic interiors of the village's inhabitants (fake intestines, fake wounds, etc.). The classificatory mechanism of the war-culture industry, meanwhile, ejects that which is outside its schema: the incommensurability of the embodied lived experience and knowledge of the role-players. As military personnel seek to balance various training imperatives, they continuously evoke the notion of "Left and Right Limits"—that is, a spatial metaphor for the parameters within which a military endeavor would be considered to be functioning in a successful manner. Trainings are a balancing act that must accommodate various operational, tactical, financial, logistical, and cultural goals. In some instances, these goals conflict: for example, a military desire for "cultural authenticity" or "fidelity" might be supplanted in some instances by a particular tactical or logistical mandate.[4] Indeed, other thinkers of military and civil defense simulations have noted an ongoing balancing act between realism and accomplishing objectives (T. Davis 2007; Magelssen 2014).[5] Meanwhile, the trainings proceeded amidst a range of military imperatives that were typically not transparent to the cultural role-players themselves. This gap generated tensions and surfeits for role-players asked to embody their cultures—but not on their own terms.

In *Being and Time*, Martin Heidegger develops his conception of the "tool." He contends that the tool is essentially "something in order to." He contrasts the tool, as existing in the state of "readiness-to-hand," a state of pure instrumentality (you don't even realize the hammer is working when it's in use), with that which is "present-at-hand," that which is subject to inquiry, an object of apprehension. According to Heidegger, the working tool is ready-to-hand, meant to remain in action seamlessly, never revealing processes, but rather only effects, and even receding from the consciousness of the user: "What is peculiar to what is initially at hand is that it withdraws, so to speak, in its character of handiness, to be really handy. What everyday association is initially busy with is not the tools themselves but the work. . . . The work bears the totality of references in which useful things are encountered" ([1953] 1996, 65).

The working tool, in this sense, is operationalized for "work." This is its sole use, which is determined from the outside, closing off other forms of being and becoming. Conversely, Heidegger proposes, it is only "when we discover its unusability, the thing becomes conspicuous" (68). As role-players negotiate the military's Left and Right limits, there is a perpetual presence of excess: excess knowledge, excess affect, excess risk, which are present *in spite of* the system's parameters. In most instances, these excesses do not upend the system; in other words, the tool is meant to sleekly work, and thereby withdraw from notice.

.   .   .   .   .

Recent anthropologists of American militarism have noted how the soldier's body and psyche have been newly constituted through trainings (MacLeish 2012; Sogn 2014; Jauregui 2015).[6] Cultural predeployment simulations likewise might be understood as mechanisms for the transmission of a cognitive and sensuous knowledge, to become sedimented within the warrior body. One soldier explained that "the idea was to make it real," adding that that knowledge would ideally come back later when it was needed: "People revert to their training in the 'oh shit' situations." The training formation is more self-conscious than Pierre Bourdieu's habitus, that "structuring structure, which organizes practices and the perception of practices" ([1984] 2010, 170). Perhaps, Marcel Mauss's "techniques of the body" offers a more apt parallel: "the action is imposed from without, from above" (1973, 4), through a paradigm of gaining tacit knowledge alongside more explicit knowledge.[7] Through the embodiment of this knowledge, the goal is for the soldier to become more acclimated within the warzone.[8] The documentary filmmaker interested in impacts of the training, and who had conducted prior interviews with training soldiers, suggested to me: "Frankly, I can see how at age eighteen, never having been in combat before, never having been out of the U.S. before, you freak!

Part of what the [military] has to do is to train that out. You've got to train out the freak-out."

That "freak-out" might be precipitated by any kind of dislocation or estrangement, cultural or otherwise. As one member of military personnel at a training explained: "By the time they leave here, they're pretty much ready to handle anything." To this end, mock villages generate sensory and cultural transport to inoculate the soldiers to new bodily experiences through a range of methods, different in each village: the call to prayer is played at the appropriate intervals; mock improvised explosive devices (IEDs) reverberate through the village; women lament after funerals or are audibly "beaten" by their husbands; I even heard anecdotally about one village where the putrid waft of mock mass graves is created, by burying rotten meat in a ditch. As the soldiers mute the impact of cultural and sensory dislocation, two major military epistemological schemas come to the fore, within two different kinds of simulations: one that relies, in my categorizations, on identification/negation and another than relies on incorporation.

The first kind of simulation, the simpler genre, includes what military personnel call, variably, "Situational Training Exercises" (STX) and "Soldiers Urban Reaction Facility" (SURF), for the space they take place in. These exercises, "short, scenario-driven, mission-oriented, limited exercise designed to train one collective task, or a group of related tasks or battle drills, through practice" in some instances act as a precursor to more complex simulations (the FTX and FSO to be described shortly).[9] Sometimes the STX/SURF simulation is set up in a quadrant, like in the opening description of this chapter; in other instances I observed, they unfold over one mock street or "lane." This stage might be typically seen as a "reconnaissance period," where the soldiers gather initial information about the mock village. Soldiers are instructed to map out and decipher the so-called human terrain, the persuasions of different sects or groups within a population. One developer of the simulations explained this initial process: "They have to determine who the players, the power-brokers are. How do they fit together in a matrix? What are their motivations, their spheres of influence? Who do we support? Who do we marginalize? Who do we ball up? Then, how do we influence the targets the way we want to?"[10] "Culture" as a tool is mobilized both as the content to understand (the social sphere) and as a hermeneutical mechanism to locate some individuals as supportable, others as marginalizable, and others as those who should be within the "kill-zone" or "balled up."

In prototypical versions of these field exercises, the soldiers move through a series of cultural and tactical dilemmas facilitated by role-players who enact interconnected characters in a village. The soldiers are typically required to ascertain what the political and/or sectarian orientations of the villagers are, and if those orientations are malleable and might be influenced. Invariably in these scenarios, some of the villagers have connections to the militias. In some instances in the simulations, even power brokers in the village, such as the mayor, the imam, and the police chief, are affiliated with militias. In reconnaissance missions of this nature, soldiers are instructed to familiarize themselves with the human terrain and then identify (and then eventually eliminate) the "bad guy."

Many of these simpler simulations function something like a mechanism of identification, where fixed categories of identity are linked to categories of imagination and action. The terms of identity function within a kind of "closed equation of representation, x = x = noty" ("not y") (Deleuze and Guattari 1987, xiii).[11] In the role-play, individuals are understood, typically, to be x (pro-American) or y (militia members). Thus, to affirm that an individual is x is to ascertain that they are consequently "not y." Within this calculus, the training soldiers should be able to predict behaviors and motivations based on allegiances.

The second and more complex and contingent form of simulation I observed seeks to offer enough cultural and tactical knowledges to make various acts of war containable by the soldier. I heard this type of simulation generally described as the Field Training Exercise (FTX), which are names for more elaborate multiday simulations, with more cause and effect and contingency.[12] The FTX presents a far less stable and knowable world than the simple simulation; however, it still posits that American soldiers can become insiders in the warzone and ultimately both the adversary and even a deeply contingent act can be tamed. As one military brochure at a conventional army base explained, the soldier must learn to anticipate what was seen as an increasingly elusive adversary: "The evolving sophistication of military equipment and technology, pitted against an evolving, ruthless enemy, demands a comprehensive battlefield that realistically simulates the complexity, tempo, range, and intensity of current and future conflicts."[13] James, the war-game theorist who earlier suggested that "any form of unpredictability can be incorporated if we do it right," explains this approach in the military. Once an object is assigned to its proper category, we might make "sense" of it, but what is more: theoretically, all forms of contingency might be incorporated within that schema, and thereby ultimately contained by the soldier in the warzone.

Nicknamed "Choose Your Own Adventure" by its creators, these war games are playgrounds for military planner imaginations—with all of its constraints and orientalism. They typically last twenty-four hours a day for approximately a week, and they attempt to incorporate all potential cultural and tactical outcomes. I spoke with a major in the Special Forces community about the principles behind this category of simulation. He explained:

> "We don't give [the trainees] the keys to the kingdom. Their job is to get the realities on the ground and to achieve adaptive thinking: it's about realities on the ground rather than their preconceived notions."

> "And how do you accomplish that?"

> "We can't spoon-feed them. There are natural consequences to their decisions. If by day three or four they're lost in the woods, we'll throw them a bone."

Before the exercises were set to begin, I went out for Mexican food in a strip mall with Jimmy, a soldier who assists in the scenario-writing process. I asked him how he enacted these goals in the scripts, and he explained:

> "When the students go through [our] course, they ask, what was I supposed to do? X led to Y. So I do Y? But all we want is for them to learn that decisions have consequences. One issue is that the students frequently ask previous training classes about the exercises to get the 'answers.'"

> "Can you give me an example?"

He explained that in a previous year's scenario, the villagers were getting ill from their water source and the new class discovered this:

"[The previous class] told them to boil the water. But, the *answer* is to ask the right questions. In this case, it would be to ask to sample the water, so you can figure out what is causing the illness. The second group boiled the water, but they didn't ask the right questions—they just used received info."

"So how did the military leadership react?"

"So, we decided to change it; i.e., to figure out what would make people sick *even* if they boil the water. So we decided the issue was arsenic poisoning. And boiling the water in this case would intensify the poison."

For the military leadership involved in creating the scenarios, this incident was indicative of a reticence to employ critical thinking and deserved a punishment: in this case, the notional death of everyone who subsequently drank the village water.

The closed equation of identification (x = x = noty) might best be counterposed to the Deleuzian equation that represents nomadic thought, the infinite accrual of successive terms in an equation: $(+ y + z + a + \ldots [+ \text{arm} + \text{brick} + \text{window} + \ldots])$.[14] This latter equation suggests a set of bodies and forces and desires and possibilities that are always moved forward by more ellipses. In more sophisticated war games, not only cause and effect are woven into the games' permutating storylines; the soldiers are also required to anticipate "second and third order effects"—the potential ripple effect of their own decision-making.[15] Moreover, in this kind of simulation, any chain of action can theoretically be predicted or interrupted. While we ate chips and salsa, Jimmy told me how in some instances, a team of simulation writers would architect the simulations deep into the night: they had to take into account all the prior decisions the training soldiers had made; how things had gone awry; and what sorts of lessons might be most effective for them at this juncture.

One afternoon, I stood with a public affairs officer in the midst of what some soldiers called "the sand box"—sprawling land where war games were staged. Within our frame of a view, a police station, a prison, a mayor's office, a bicycle repair shop, an open-air market, a sprawl of goat pens, a Sunni mosque and a Shia mosque, a dentist's office, a thicket of streets with laundry lines and a scatter of children's toys. A role-player was selling lettuce in front of a white door marked with one rust-red handprint. The tea-man was pouring tea. Women sold flowers. Several American soldiers on the edge of the street and for the moment ignored by the villagers, were looking for Improvised Explosive Devices (IEDs). Amidst this formidable buzz of activity, I asked the public affairs officer to describe some of the more complicated storylines to me, and how the military handled the twists, turns, and contingencies of everyday life. He described to me a scenario generated at one of the mock villages in reply, offering a set of conditionals and their unfolding outcomes. If, in a scenario, a black marketer meets with a criminal entity to buy stolen fertilizer, transport it to another town, and sell it to insurgent forces, then the plant manager must realize the fertilizer is missing, report that to the Iraqi police, and the information should eventually get to the training soldiers, who must immediately react, by, for example erecting traffic control points or investigations. The trainees can catch the black marketers and detain them. If they are equipped with the proper knowledge, distilled into operationally useful information, any "bad guy" or "bad event" can be averted. This same logic was explained in another recent monograph about the war trainings: a public affairs officer explained: "If this was done perfectly, the Americans should have been able to decipher the puzzle, go to a village, 'knock on door number three and find the leader of al-Qaeda" (Magelssen 2014, 160).[16]

.   .   .   .   .

Postcolonial thinkers (Said 1977; Bhabha 1991; Appadurai 1996; Trouillot 1991)[17] and anthropologists (Clifford 1988; Rosaldo 1993; Chatterjee 1993; Skurski 1994)[18] have long critiqued synchronic, essentializing representations of non-Western Others that were characteristic during colonialism, and which have been mobilized dangerously in wartime (Barkawi and Stanski 2013). Contemporary military epistemologies are saturated in long discredited logics that emphasize essences, origins, and authenticity. Within the training simulations, military personnel sought to stabilize cultural knowledge within containable forms of space and time. One hot summer afternoon, a military contractor hired to build sets and run simulations invited me to a prop warehouse. While I waded through "bloodied" knives and shimmering Middle Eastern cloth in the enclosed woody-smelling heat, he described his company's philosophy:

"You do the best you can to make it look the part. But the *inside*, what the role-players are trying to replicate—language, culture, norms—that's the most important."

"And do these performances, this *inside*, happen organically, on their own?

"Over time, the role-players are further away from being in-country. People come to us and they are trying to assimilate here, and we want them to act like they haven't assimilated. You have guys like [our Cultural Advisor] to take them *back*."

The military contractor suggests that the Iraqi role-players' recent time spent in the United States may potentially corrupt a purer and more authentic relationship to their own communal identity. However, he also thereby effaces globalization and the complex amalgam of Iraqi culture and history, as well as the impact of its most recent entanglements: the American incursion itself into Iraqi borders—indeed, the mark of his own presence. In military simulations, role-players are asked to act as exemplars of their cultures, without a determining agency in where culture begins and ends: Iraqis, as it were, in a Box.

Yet: the presence of actual Iraqis was simultaneously epistemologically essential within the military project. American military leadership and training soldiers frequently emphasized the importance of learning about Middle Eastern culture as a component of their training. When asked why Middle Eastern role-players were necessary in this pursuit, military personnel described those individuals as a metonymic access point to the region: "The role-players are a whole group of Arabs in order to understand the Arab world at large." A captain helping his team train elaborated: "The role-players are *just playing themselves! They're just being a different culture!*" Indeed, in a regional studies course I observed, training soldiers were instructed to my astonishment that "being" was a particular property of Middle Eastern cultures. Although one instructor disputed this approach, explaining it was outdated, another instructor explained: "It's useful to think of it this way: there are 'to be' versus 'to do' cultures. 'To be' cultures are driven by who your family is, your tribe; honor and identity are derived from this." He explained that Western cultures, to the contrary, were more constituted via *doing:* creating oneself as an individual in the world through education and work. After the class, I asked the training soldiers what they thought of this breakdown between East and West. One offered: "The to be/to do stuff is fine . . . If a major is down-range and is, like, dudes, you guys are 'to-do' [the individual] and they're 'to-be' [family], that's still something useful."

In this constellation, role-players became embodied technologies of *being-Arab,* enacting archetypes, rather than of *knowing or translating* anything particular about the Middle East.[19] Meanwhile, the backgrounds and experiences of Iraqi wartime intermediaries—urban, educated, with particular interests in the West, and each with varying motivations for working with the military—render them even *particularly* unsuitable to embody archetypes that they would find personally resonant. Far from existing as an Iraqi-American military heterotopia, complex power relations marked mock villages as role-players were used as cultural technologies within military terms.[20]

While role-players were assigned an ontological primacy by the military, many role-players described the ambiguity of their epistemological position—proprietors of culture, but not on their own terms. Although role-players typically understood themselves to be subject matter experts, and the sole ultimate bearers of authentic cultural knowledge, extremely granular cultural detail was often exterior to the parameters of other (i.e., operational, tactical, logistical, financial) military training objectives. In such an environment, there can be surprising and seemingly contradictory outcomes.

One role-player, Khaled, explained: "Sometimes [the soldiers] don't listen to me. You're trying to teach them culture, and you have no say in it. It's mine. I'm here for you. Not for myself." Another role-player, "Salem," described culture as a product they were offering, and expressed frustration when the roles and scenarios they were given by the military impaired that product: "We have to provide service the right way. You have a customer. The customer wants it perfect. Why is a Kurd role-playing an Arab? Even his Arabic language is not correct."

In another instance, Iraqi role-players were hired to portray Afghanistan, an instance of compromised authenticity that startled the role-players. Because military contractors could not find enough Afghan contractors in time and did not want to delay the exercise, they chose to hire Iraqis. The military personnel involved told me that the soldiers didn't have language skills at that stage in their training and were using interpreters in the training. Thus, even though the role-players were speaking Arabic rather than Pashto or Dari, the training could still go on. In this model, language appeared to constitute little more than disorienting noise. In this ongoing paradox, while military personnel emphasized the importance of the authentic core of culture, that culture could readily be substituted—when necessary—for another, with an entirely incommensurable history.

Amidst a wide range of training imperatives, from cultural to tactical, military supervisors regularly noted the importance of keeping all aspects of the simulations within objectives, or in military lingo, "Left and Right limits." One of my military interlocutors who helped to devise scripts explained: "Role-playing is an effective training tool, as long as it's used appropriately and in accordance with the training objectives and doesn't get off to something crazy, which happens." In order to help the role-players hew to military objectives, they are coached by contractors and given preparatory materials that differ in scope, according to the objectives of the exercise. Role-players often received a "country or village study," with background on the military and political situation of the fake (Middle East inflected) world they are about to enact. Information was sometimes mediated through another role-player appointed to explain the scenario rather than distributed to the entire group. The specificity of the "scripts" varied. Role-players with more basic roles were often given their name, tribe name, place of origin, profession, and their sympathies (pro-American; anti-American, etc.), and sometimes a list of things they are to ask the American military for (water, medicine, etc.). Those cast as key leaders, such as mayor or imam, typically received considerably more detailed scripts, with specific lines.

These directives continuously reminded role-players that they in fact are not simply "being-themselves," but rather embodying specific constructions of their countries of origin—all of which are circumscribed by military imagination and fantasy. Military leadership often emphasized the importance that role-players cleave to military objectives. Prior to one simulation I observed, the military contractor overseeing the role-players announced to them: "This should be real but *controlled*. Not spontaneous ad-libbing. We can turn up the heat and complexity, but that must be controlled." In other instances, they were encouraged to improvise. Their employer told them: "Whatever you say is yours. However you want to approach the situation is yours. In this scenario, your goat is stolen. You want to yell, then yell." However, in each instance, the role-players were reminded to reorient themselves around the soldiers' learning objectives if they went too far. A scenario writer at one military base who was especially interested in creativity within the scenarios noted that sometimes the role-players went too far off track and had to be reined in, moments he called "red lemmings"—tumbling off the military edge.

Several majors indicated their inclination for the most predictable and controllable mechanism available to them, expressing a preference for role-players culled from U.S. military manpower rather than the more recently hired Iraqis, who were known as the "cultural role-players." These majors explained that U.S. soldiers, especially very experienced retirees, were more attuned to military goals than the Iraqis: the soldiers "are so good you tell them what you want to see pulled out of them and they give it to you." Even amidst cultural objectives and an interest in creating an "authentic" environment, in some instances, this gave way to the military desire for the trainings to proceed as much like a machine as possible. That is, ultimately an ideal performance was when a role-player "pulls out" only what is specified, without adding anything irrelevant.[21] However, rather ironically, as military personnel work to become cultural insiders, the inner lives of their cultural interlocutors were often deemed irrelevant or distracting.

Meanwhile, as I will further discuss later in this chapter, most role-players, who were regularly frustrated when their expertise was disregarded, conversely focused on image management in order to retain their work, making the simulation run smoothly and "look good" even as they recognized inauthenticity—with little choice but to work as compliant military tools.

·   ·   ·   ·   ·

The working tool, in Heidegger's understanding, is meant to remain seamlessly in action, never revealing processes, but rather only effects. Conversely, when a tool goes awry and cannot be used as desired, it at last becomes wholly conspicuous to its user. In his discussion of Heidegger's notion of the broken tool, Graham Harman offers an intriguing analysis: "The broken tool counts as the first way in which an entity is freed from its contexture, released from the dimension of reference. Here the tool is encountered *as* tool rather than quietly functioning one. Fractured equipment emerges as a determinate entity, torn loose from the totality; to this extent, it attains a kind of presence *in spite* of the system that tries to consume it" (2002, 49).

Harman's analysis is perhaps too far-reaching here: when the tool breaks and becomes present-at-hand—available to perception to the user—perhaps it still breaks *as a tool*. It attains a momentary presence perhaps *within* the system rather than *in spite of it*, not truly liberated from use. For role-players laboring within a military system seeking to keep their jobs, acquiring any such visible presence was not an available option. As they engaged with the military system and its representations of the Middle East, role-players regularly wrestled with uneasiness and an experience of holding excess knowledge with no outlet. Amongst themselves, they often rejected military parameters. Nonetheless, they still hewed to them carefully in order to keep their jobs and only occasionally spoke up. That is: they generally functioned as *working* cultural tools within a military system, making the game (at least appear to) run smoothly while still feeling ongoing and typically unexpressed disjunctures within the simulations.

I asked several role-players how they felt about how military parameters functioned within the simulations. "Salma" spoke of the frustration provoked by the confines of her role and how it did not permit her to offer the scope of her knowledge:

> "I had so many things I wanted to say to them, and I couldn't! I wanted to say: there are no services—no electricity, no water; there are explosions and we are afraid."
>
> "How could the role-play have been better?"
>
> "If you have experience, you should be able to say it! I have information, and I wanted to bring it to them."

"Yusef" cautioned that other role-players should not diverge from the script:

"Sometimes you can add a little, but be careful. If you add one word, you could change the whole scenario."

When role-players found scenarios incorrect or artificial, they often kept silent:

"I wouldn't tell them they're doing something wrong. I would *cover* for them if there's a mistake. That's why they promoted me to sergeant in the scenario."

Hussein added:

"No one speaks up. No one stands up for anyone but themselves. We want to keep our jobs."

Occasionally role-players would speak up; however, this more often occurred in the less stringent simulations I observed. Once, Lubna, a role-player distressed by the inauthenticity of the simulations, complained to a soldier, noting that the setting was incorrectly rendered and that she didn't understand why people of different nationalities were role-playing within one mock country.[22] She explained:

"There is no resemblance. Even the rocks are different. This place is full of Egyptians. Why don't they make a better resemblance? I was angry about this, and I told the captain."

In some instances, a role-player was appointed a "cultural advisor" and in certain circumstances the military and/or the contractor welcomed interventions into the simulations. The individual with this title typically had a range of roles, from procuring props for the role-plays to having extra sway in the hiring of new role-players. One individual with this job, when asked if he relayed the other role-players' suggestions to his boss explained: "I take ideas. I go with it if it's right. I have to confirm with the army." In occasional cases, the cultural advisor would become an informal and deeply trusted advisor to the U.S. military leadership, who was urged to intervene regularly when he believed the scenarios were incorrect. I spoke at length with one such cultural advisor, "Ghaith," who sought to improve the scenarios within the simulations whenever he could. He recounted:

"In one scenario, there was a fake election. An elderly role-player walked in and they [the soldiers] didn't search him. I asked the soldier, why didn't you search him? The soldier said, 'They told us not to because he is older and needs respect.'"

"What was your response to this?"

"I told him: 'I don't give a damn about culture when it comes to safety. Does this guy have a [suicide] belt on him?'"

Ghaith explained that he felt affirmed in his correction after a conversation he had with that same soldier who later returned from a deployment: "The soldier told me he benefited so much from [the simulation]. He was in Iraq for nine months right afterwards." Ghaith was confident in the utility of his cultural knowledge, emphasizing the importance of the training in preventing the soldiers from making grave mistakes:

"In each [training], the commander must sign off if a unit is ready or not to deploy. Sometimes they have to repeat the training. This stuff is not a joke."

Ghaith was proud that he had a role in improving those scenarios:

"The scenarios were not accurate; that's why we were there."

In Ghaith's opinion, the military wanted to make the scenarios more authentic, but the contractors they hired did not want to put in the work to change them once they were already written:

"The military agreed [with me], but the contractors were lazy and didn't want to do their job. The American contractors are only after money."

·   ·   ·   ·   ·

Although Ghaith had the ear of an American commander, his position was rare. At one military base I visited, which had an Afghan mock village, "Bill," my military escort, asked me if I wanted to speak with any of the Middle Eastern role-players, as this might "require negotiating some red tape." Later that day, Bill brought out two role-players appointed to speak with me. The three of us sat at a table and the military escort sat on an adjacent ledge to listen to the conversation. All the role-players in the

Afghan village were Dari, Farsi, and Pashto speakers, languages that I do not speak. In contrast to my conversations with Iraqi role-players that took place in Arabic, I interviewed these role-players in English. I began to ask them a series of generic questions about their backgrounds and their motivations for doing the work. One of the role-players, a middle-aged woman named Fatima, replied:

> "I do the work to be helpful. [The U.S. military is] going to my country, I want them to go to the right place and do the right work. They might know about Iraq, but they don't know about the [Afghan] culture. It is good to help them."

The other role-player, Amir added:

> "The American military doesn't know anything about the Middle East. We're explaining the region to them."

In each instance, the role-players who were under observation offered a motivational logic akin to military discourses about the importance of role-players and cultural awareness, lauding the importance of cultural awareness and sensitivity. Although a number of the role-players earnestly support the American military and are seeking to become American citizens, for many interviewed in other contexts there are complicating factors, which make for some degree of ambivalence about the job. However, in the context of an interview observed by military staff, a more seamless articulation of motivation was offered. Yet, this brochure-worthy performance was interrupted when I was given a sudden card within the game: when Amir shared his biography, he noted that he lived for a number of years in Kuwait, and we immediately began to chat in Arabic. Bill looked up from the ledge, taking note, but he did not comment.

Thereafter, I resumed the interview in English. At this juncture in my research, I had already conducted extensive interviews with Iraqi role-players, and I was thus aware of which turf might be challenging for my interlocutors. Asking about their family and their countrymen's reactions to their work with the American military often brought forth moments of internal ambivalence and contradiction. I thus asked:

> "Do your family and friends in Afghanistan know you do this work? If not, how would they react?"

Fatima reacted with an uneasy affect to my question and after a pause, replied:

> "It isn't a problem. It is important for me to follow my own opinion. I don't care what anybody else thinks. I am helping my country. I have citizenship here and this is my country now."

Rather than probing further at this juncture, I asked Fatima how she felt within the simulations. She replied:

> "I feel that it is very important for the troops to help find the bad guys."

I next introduced moral ambiguity into the rubric, asking how the American military can truly determine who is "bad" and who is "good," if sometimes people work for insurgents because the lives of their families are threatened and sometimes people take money from militias because their families are starving. Bill craned up in his seat to listen more carefully to their replies. Fatima was carefully watching Bill on his ledge as she answered, reinstating a calculus of those who were authentically "good" versus those who were authentically "bad":

> "You need real friends who won't trick you. You need to get information from honest people."

Amir interrupted her and reintroduced a more contingent logic, wherein it might be less clear who the "bad guy" is:

> "You have to find work for people who have no work; or else they will work for anyone."

The small group dynamic soon became more complex. Fatima was performing the military line; Amir was pushing against it. Amir then whispered to me in Arabic:

> "Fatima was not telling you the truth before at all. You cannot tell anyone over there [in the Middle East] about doing this work or they will be angry. She is ashamed to admit that others would judge her for this. Most of them don't do this work genuinely. They live a contradiction. They are loyal to their first country, not to America."

Later in the day, after the interview was over, but within earshot of other role-players and as well as military personnel, Amir added in English:

> "They need to hire role-players with loyalty to this country. I'm not scared of anyone. I talk. I'm free. I can talk about religion and the bad side of this culture. I don't care. The army pays for you: why don't you share your experiences with them? I feel a responsibility to do so. People hide information because of shame. We should help [the military] learn about what's bad in our culture so they can cope with difficult situations."

For Amir, other role-players were reticent to share additional knowledges not only out of fear of exceeding military parameters and potentially losing their jobs, but also out of fear of betraying their own allegiances. Nonetheless, the following day, Bill described Amir's speech as an "incident":

> "If those idiots run off the mouth and speak their emotions, the companies they contract for could hear about it and fire them."

He then cordoned me off with a group of several other role-players, who had presumably been more carefully primed for the conversation. The interview proceeded in an airtight manner, as the role-players described their commitment to the American military's goals. Without negating the attachment of many of the role-players to their new country, this incident surely provides a portal into one limit-point of role-players' epistemological maneuvers. In order to function as cultural training tools and be unnoticeable to their users, role-players were required first to inhabit fixed archetypes that in some cases did not ring true to them. Second, they were required to seamlessly wash over the contradictions in their own particular positions: essentially, although they were hired to embody versions of their cultural selves, their own precise location *in relation* to that work had to be suppressed in order to make the simulation work smoothly/conform to expectation/look good: that location, riddled with contradictions, accusation, and in some cases great ambivalence was a source of potential friction. The role-players might alternately theatricalize adversaries or allies or informants, but only on military terms.

.   .   .   .   .

Indigo smoke rises behind the houses of the mock village, an obscuring haze through the pines, to confuse the adversaries in a nearby simulation. A group of soldiers congregates to investigate a house, which they suspect might harbor the militia. Fatima and her daughter Leila wait inside the house. Fatima's son, Omar, is a suspected militant. The female soldiers, wearing veils, knock on the door and speak carefully and deferentially to Fatima and Leila as a means of establishing rapport. The soldiers then asked about Omar's activities and social connections, and if the women had a picture of him. Fatima and Leila initially said she didn't know anything about the topic and did not have a picture of him. But as the soldiers probed, Fatima eventually acquiesced, offering that Omar had become friends with some "bad people." She told them that she could provide no further information. The soldiers departed, equipped with the information that the son had "bad friends" and was thus a "bad guy."

In the scenario, Fatima and Leila, who yield to the American soldiers, are the x, akin to the self or ally; the son is y, otherwise known as the adversary. However, a much more complex scene also unfolds. Fatima and Leila have been transformed, not only to American allies, but also into informants to the military and traitors to Omar, their own brother. The subsequent scene both briefly acknowledges that there is a more multilayered transformation, from villager to U.S. ally and informant—and one with terrible stakes for the local individual involved—and occludes what might be the cause of the incident.

After the soldiers finished the interrogation and departed, the house was briefly still. The videographer flashed a grin at the role-players playing Fatima and Leila, whispering: "Are you ready for your debut? Tell him to be soft!" Leila, swaddled in a heavy veil and long dress, held a small bottle of fake blood. Out the window of the flimsy house, I saw Omar, dressed entirely in black, indicating his militia affiliation. Right before he lunged through the doorway, Leila tipped the fake blood in droplets all over her face. To punish her for affiliating with the American soldiers (that is, for acting as x to his y), Omar simulated throwing Leila in the corner, into an inset in the wall, then pounding her with fury, while she crumpled into harrowing screams. Fatima immediately threw her arms around Leila, who was also beginning to convulse with tears, and Omar dashed out.

When the American soldiers heard the ruckus, they faced a classic cultural dilemma often reprised in the simulations. The soldiers had to choose to either do something ("save her"—"men [in this case, of varying races] saving brown women from brown men") or not do something ("live and let live; this is their culture").[23] There were rarely uniform answers on the correct course of action, as military understandings of culture pivoted back and forth between standing for a more public notion (for example, the mapping of tribal genealogies or sectarian affinities) and a more private notion (a domestic sphere, an intrinsic pure zone that should ideally be left without interference). If a family member was beating a woman, like in this example, the soldiers typically heeded the lessons of cultural relativism: they had learned in their cultural training classes that an outsider intervening within the private spaces of "culture" indicated a lack of cultural sensitivity and understanding. Military personnel focused typically on the private ("live and let live") aspect of culture, until it coincided with the more public aspect (political and sectarian affiliations), prompting intervention.

Most crucially, however, the military's notion of culture is timeless and authentic and erases historical and geopolitical circumstance. Within this logic, *they* can be expected to beat their women and join militias, due to their cultures. Yet to the contrary, Omar is beating Leila because she has been coerced by the American military into risking her life to become an informant and betraying her family. The incident is thus, in fact, not a representation or act of culture in the sense the military would have it, but rather provoked entirely by exterior circumstances—an act precipitated by American occupation and war.

In each iteration of the role-play, the new group of training American soldiers reacted in variable ways: occasionally they would pull Omar off Leila; sometimes, when paralyzed by the question of "culture," they would do nothing, then hand Fatima a roll of gauze afterwards; and in other instances, they would call in local police forces to intervene. Irrespective, as the soldiers sought to decipher and lure social actors over to the American side, as they sought to understand the stakes of "culture," they continuously effaced the complex impact of their own wartime imprint.

This role-play, moreover, reveals another dark tension and excision. The scenario generates a meta-enactment of the actual interstitial position to Middle Eastern role-players. Despite having chosen their wartime work for ideological reasons and harboring powerful hopes of helping to transform their countries, they were often accused of being informants to the American military and were pursued by militias. Wartime intermediaries, who risk their lives for the American polity, become a sacrifice imagined as necessary within a rearrangement of the international order: a replicable, justifiable expenditures of life that enable the American soldier to assimilate cultural complexities on the ground in wartime. In the role-play described earlier, Fatima and Leila actually theatricalize their own expendability.

Interestingly, the role-play incorporates the perils of Fatima and Leila's position: the training soldiers witness Omar's retribution. Then, however, they willingly pursue practices that lead to its reproduction. Meanwhile the lived subject position of the actual individual who works as a role-player (psychologically imperiled by their countrymen's judgment, if no longer physically endangered for a choice of wartime work) is effaced. In the prior interview with Fatima that was observed by American military personnel, she described her motivations and allegiances within military terms. Amir interjected in Arabic, telling me that to the contrary, Fatima—and many of the role-players—inhabited a complex and sometimes ambivalent position. Like the characters they role-played in the scenario, these mediating individuals are subject to the scrutiny of their countrymen for their wartime choices. Like those characters in the scenarios, many of the role-players dodged their own deaths if they previously worked as interpreters in the warzone. In this small parable, even though Fatima acts out the informant in military terms within a military role-play, her silence when asked about her *actual* lived position in that regard mimics in a minute way the sacrificial offering of the cultural intermediary to the military project.

.   .   .   .   .

I am dressed in a long robe and a hijab that my Iraqi friends have tied securely to prevent stray tendrils of hair from coming loose. The simulation of the day is focused on language practice, in contrast to the more typical simulations that are focused on tactics and extended cultural dilemmas, wherein interpreters are used. We are told to speak Modern Standard

Arabic with the soldiers and to engage in basic conversation, entailing more free linguistic play than in most simulations. We are given a description of the fake Middle Eastern country that had been occupied by the American military and the basic storyline: we are meant to interact with the soldiers in formal (Modern Standard) Arabic and discuss the needs of our village. I am cast as an ambitious young woman who wants to study abroad and is seeking information on how to further my studies from the soldiers.

This is an opportunity, I decide, to test the "Left and Right limits." Although religion and politics are regularly woven into the scripts in pre-curated forms (the object of many of the simulations was to determine the adversary), role-players are always told to "never bring religion or politics into the box" *on their own*. This prohibition generally makes the boundary between inside and outside the box, and the contrast in behavior in each space, all the more stark. A military contractor told me that this was because the soldiers were being videotaped and should not be prompted to express political opinions while in uniform on video—an explanation that bespoke military interest in discipline and indemnification. However, several role-players conversely told me that in early days of role-playing, fistfights had broken out between role-players provoked by religion and politics, and thus these topics were taboo.[24] The latter explanation shows another military othering of role-players: in their assumption about their potential lack of control should things get "too cultural."[25]

During the role-play, I decide to bring into conversation one religious and one political topic. I first ask the soldiers if there are Muslims in America, and are they religious? Salem, the Iraqi cultural advisor, hisses "Cut," whispering to me in Arabic, "*ṭāʾfī*" (sectarian), even though nothing in my query had actually evoked sectarianism. He then summons me to the side, explaining: "There is no religion in the scenario."

As a visiting student rather than a contractor dependent on this work for pay, I am in a position to continue my experiment by feigning a misunderstanding about the game's rules. I return to the simulation and reenter into conversation with the training soldiers. I next ask a soldier why the U.S. and Israel could have nuclear weapons but no other countries could. Salem gives me a strained look from the corner, apparently baffled and distressed by my inability to follow protocol. He calls me out of the role-play: "No politics in the scenario."

At this point, everyone takes a brief break, and the role-players caution me never to bring up "real topics" in the simulations. Thereafter, still on break, I sit on the steps of the collapsible home with a role-player called Mahdi. As is the usual pattern, we immediately begin to discuss religion and politics. He recounts to me an experience where religio-politics, in the form of a sectarian death squad, nearly cost him his life. He was on a bus in Baghdad, and a militia stopped the bus and asked everyone their sect: "My friend was standing in front of me. He didn't know the right answer to the question. He replied Shia. But it was the wrong answer: the men shot him." Mahdi, who has a classically Shia name, but is in fact half-Shia and half-Sunni, answered that he was Sunni. When they asked why he had a Shia name, he replied: "Well, I really am Sunni." He recounts: "Then they said, well you're white—noting that Sunnis from Baghdad were often pale complexioned, while Shia from the south often had darker skin—so we won't kill you."

After this harrowing story, Mahdi and I return inside for the "Cultural Lunch," which functions as an ongoing portion of the simulation. As a female role-player, I spoon the soldiers' couscous and salad. We then sit in a circle on the floor and volley benign questions in formal Arabic to the soldiers, about their likes and dislikes. The surface of the scenario is taut, a translated text neutered of danger. Mahdi offers to the circle, in formal Arabic: *Hal ta'jibūn al-'anib* (Do you like grapes)? Sitting inside the theater of war, one must pan out, dizzyingly, upon the whole of the frame. This banality and controlled discourse within coexists with the seething intensity on the stairs outside. The cultural tool complies, performing the epistemological terms of its own existence. Yet meanwhile, there is a perpetual and often tacit disjuncture within the simulations themselves: military portraits of the Middle East stand vacant of those who represent them.

On another afternoon, I role-play an American NGO worker within a more elaborate and contingent simulation. I enter the woods, where several trainings of different units of soldiers are occurring simultaneously. Whorls of chemical yellow and green rise over the woods, used to provide the soldiers cover and to obscure their movements from the adversary. I walk past the pens of chickens and goats, into the strains of Arabic music floating out of the Muslim village. I note the mayor, the imam, the cluster of village women, all with consequence-bearing affiliations and attitudes that the training soldiers are required to decipher.

A military scenario writer, who everyone lauds as the puppeteer and the creative genius behind the unfolding training exercises, has cast me as an "idealistic yet foolish Arabophile American student and NGO employee who thrusts herself in a warzone to aid the displaced refugees." Through this casting move, I have gotten a sudden glimpse into how I am perceived on the base. I am instructed to be as well-intentioned yet as ignorant as possible when inhabiting my role, distributing, for example, expired packages of food and vaccines. They want me to move between the multiple mock villages, advocating for the refugees as well as behaving in a generally disruptive way. Because I speak both Arabic and French, they decide that a further dimension of my role will be as the leak between the Muslim (Arabic-speaking) and Christian (French-speaking) village. I am told that while most participants in the game are cautious, attempting to keep information close, that I should fling it everywhere, thereby stirring up conflict between the villages.

Through the course of the simulation, the trainees must discover that I am not who I seem to be, a typical feature of the more contingent simulations. Although I am initially cast as an idealistic idiot, and my information leak seems to be inadvertent, the following day, the scenario writers tell me that I was actually an undercover American agent who engages in a "honeypot operation," sleeping with the suspected head of an illegal organ-trading ring to extract information about his activities. It is my impression that this was a shift to my character on day two, and not part of her original profile. In more contingent simulations, characters might be altered or given additional layers to turn the plot.

I enter the simulation, looking for the refugees I am trying to aid in the storyline. The military contractors have positioned me within a giant fenced-in field, and there I wander. A team of soldiers arrives and asks me to open the gate so they can ask me some questions. I avert eye contact and instead run away, having been told by military personnel to not make it too easy for the training soldiers. With a small effort, they all get inside the fence. The group of soldiers chases me through the field, and they soon catch me. They ask whom I work for and what my goals in the warzone were. They ask what I can tell them. Namely, they want to know: am I of use, and why was I put in this story?

When I yield very little in response to their initial queries, they try to initiate a chat with me: "Where are you from in America?" I answer them. One replies: "Oh, I love DC. What school did you go to? What did you study?" And I still yield very little, resisting their courtship. They had been instructed to secure rapport with me, as with all the role-players, in order to extract useful information. But I do not trust them. I do not know why they have entered this fictional country at America's bidding, and the character I play is not sure if she agrees. She is reluctant, but the cards are still out; she may help them further their hold on this occupied fantasy-land. Meanwhile, I, Nomi Stone, do not agree.

The simulation writers initially asked if they could use my real name for my character. One of the scenario writers offered: "Let's just call you Nomi Stone in there!" He wanted to add a layer of oddity and complexity, another kind of decipherment to the game, for the training soldiers, explaining: "If the [soldiers] are smart, they will google you and find out you're a poet (huh?) and that you're Jewish. This will make you especially confusing to them . . . a Jewish Arabophile???"

The request to assimilate my actual biography into that apparatus further reminded me of the strange work of the contingent machine: it is a nomad, with its quick mimicry and appropriation of the real. That machine tries to incorporate the totality, so that the soldier might be able to predict and contain the imaginable set and become insiders within it. Although these more contingent simulations tried to bring in "real" experiences of wartime, as much as possible, I hadn't before witnessed the use of actual names and identities woven in—and it is possible that as a visitor to the war games, a student and nonlaboring exception, that this suggestion was bestowed upon me, almost playfully. It was also not my impression that there was a practice around deliberately confusing simulated and "real" identities—but perhaps this might have foreshadowed when I played "Gypsy," another character I will describe in chapter 6.

I immediately rejected their idea to use my name, which felt unnervingly, intimately intrusive. I did not want who I am in the world turned into a character, or worse, an information module. Meanwhile, the Nomi Stone they would adopt in the scenario had politics that were not my politics.

How could I reconcile myself to this fraudulent incorporation of who I am? The scenario writer's proposal unnerved me in particular as I remembered the estranging labor that the role-players undertook every day. They were asked to be-themselves, but within the military idiom. As an unpaid American visitor (but funded by a university), rather than a Middle Eastern worker (within a capitalist system), I was afforded a range of privileges that were unimaginable to the Middle Eastern role-players. They were not offered a chance to dispute the characters they had to inhabit; they were not permitted political complexity. I have suggested here that the role-players were not able to perform Arabness *"on their own terms."* This surely begs the question first of how many employees ever enact labors on their own terms, as capitalism itself—and the military-industrial complex certainly—proposes other terms from the outset. The question is perhaps all the starker when that labor consists of acting out a (cultured) version of the self but a hijacked version. Indeed, whether any individual can have "their own terms" at all amidst vaster, powerful, and often naturalized (political, economic, cultural, etc.) structures is a more existential question embedded here. Nonetheless, some might maneuver these structures more easily than others. The military planners seemed a bit surprised by my resistance, but they acceded to my request without objection.

The military paradigm in question relies on a double movement: on the one hand, the training soldier is meant to be transformed: his or her body and its surrounding technologies are enhanced and his or her perception and preparedness are enlarged. Yet in order to achieve this end for the soldier, the role-player takes on an inverse property: his or her subjectivity, perception, and experiences are restricted and sometimes even foreclosed altogether. When a person is turned into a tool, this forecloses other forms of being and becoming. For Heidegger, the tool becomes visible when it becomes unusable to its user, when it in some way breaks—in this instance, when subjectivities rise to the surface. I was the only one positioned to externalize that the tool is broken; indeed, in an ideal world, existentially speaking, the tool is no such tool: human beings are more than repositories of use. I was the only role-player that day who had the luxury and privilege to refuse entry into that particular set—even as I was bound to other rules of the system. "No," I said, disturbing the military plan, "you cannot use my name."

[Field Poem Fragment]

Crying Room is a booklet that tells how
a body is undone by plot.

[Field Poem]: Driving Out of the Woods to the Motel

[Character 1, Game 2, puts the keys in the ignition]

Get gas. Get a Coke. Drive past American houses, shut in a chokeweed
of chickens and trash. To the highway and turn the radio up. Make
a to-do list. Tuition, groceries, pin number, food stamps, and your
cousins wait on Skype. You could get another job. You could make your
wife proud. But the Walmart application has 70 questions. Does
Walmart think it is the White House itself? For example, the unhappy
customer comes up to you and says: Walmart has bad service. Do you:
A) Give the customer a form to lodge a complaint? B) Tell the customer
that Walmart is understaffed today and apologize? C) Apologize and
say Walmart is doing its best? Save money, live better. Be a checker
at the grocery. Go to the center of the weather, be a good father, be a good
daughter, honor your mother, call her on Skype. You are Moe. You are
Joe. You are Raki, for short. You were the best damn interpreter
the soldiers ever saw. They trusted no one, but they trusted you. Live with
honor, live in God's hands. But why did they hire the other guy? Why
did your child almost die but

> not die when the bomb sheared this
> sky upon them?

5 Affective Maneuvers

Over a tiny bodiless coffin in an isolated American wood, Iraqi role-players cloaked in black are wailing to a crescendo. In a cultural training simulation some military contractors call the "Crying Room," the women congregate around a small wooden coffin decorated with a bouquet of fake flowers, a framed picture of a child ("the deceased"), and a black hijab-cloaked Barbie, for regional flair. The women have howled with grief dozens of times, performing for three days straight, for each new rotation of training U.S. soldiers. Their voices are hoarse. "I sound like a donkey," one of the women laughs, mimicking a braying noise. The simulation always starts similarly: the Qu'ranic recitation begins to play from a boom-box; the role-players, clad in embroidered black dresses and veils much like the doll's, wail and hit their thighs and the tops of their heads in synchronicity. Once, when they are testing the music, a military contractor walks in the room and playfully begins to dance to the reedy lilt of the Qu'ran, but one of the role-players chastises him: "That's God music!"

In the scenario, the U.S. soldiers had inadvertently crushed the child with their tank and now must pay their condolences, while ideally acquiring intelligence about the family's connections to the militia. When each unit of soldiers enters a cultural simulation, the first step in the training program is to establish "rapport" by assuming sympathetic postures and mirroring their interlocutors while their interpreters translate; in this particular room, they are often flustered by the cacophony of Arabic and weeping and seek to regain their composure.

The performance of affect within role-plays is subject to military prescription and persuasion. Yet, whereas military personnel largely curtail or even efface role-players' epistemological excesses in favor of more rigid cultural translations, charged affective surpluses take on a range of other contours—encouraging some intensities as registers of the authentic and dissuading others as charged distractions. For role-players, affective performances are typically described as fatiguing. But in some instances, affect seems to *act,* jolting into a controlled field, and offering a changed experience for the person experiencing it.

Similarly to military interests in "authentic" embodiment of cultural knowledge, affective "authenticity" is encouraged, a useful atmospheric backdrop or narrative movement. Military personnel and contractors ask role-players to perform storylines as authentically as possible, to transmit the intensities of war to the training soldiers as well as to allow the soldiers to practice the craft of rapport-making and information-extraction in various heightened circumstances. At one exercise, a major explained to a group of role-players the importance of performing with authenticity and intensity:

> "The villagers need to be cognizant of every activity [that is] taking place. If there is indirect fire in the village, you cannot play cards or soccer, as if nothing is going on. You have to make it real. Women should be scared. Men should be scared."

Role-players frequently would discuss the importance of making their roles "real," on an affective level, as essential to the job. One role-player explained: "Even if it tires me out, I have to make it as real as I can for the military. I have to cry if they need to me to cry." I myself role-played several times over the course of my fieldwork, and I was instructed by military personnel to make my enactment as emotionally convincing as possible. In particular, when I role-played a guerilla fighter, I was instructed to tell a heart-wrenching story about the death of my husband and child while in role. Before I went out, military personnel told me: "If you can work up some tears when you tell your story, that would be good."

Indeed, certain kinds of intense performances are seen as especially authentic and applauded by military employers, and some role-players assert that such performances make it likelier for them to be rehired. In fact, at one military base where I observed exercises, military employers gave role-players military coins to honor them when they performed with particular intensity. One role-player, "Lubna" narrated one such incident: "Once I got everyone crying. I told them, pretend it's someone you love. We began to cry, all of us. 'Dahlia' began to cry and couldn't stop. They gave her a special coin to thank her." While military coins create an economy of honor and gratitude, the moment also explicitly inscribes affective performances into a commodity logic.

Although military personnel and contractors emphasized the importance of intensity and authenticity, it was also essential that the affects of the role-players be circumscribed within appropriate, controllable, and operationally useful parameters. A major had cautioned the role-players: "We can turn up the heat and complexity, but that must be controlled." Further, although military contractors encourage affective intensities in the "safe" expression of grief, they interrupt them if they move into an unscripted and volatile expression of rage. While Dahlia cried earnestly and couldn't stop and was honored, moments of apparently authentic anger were not likewise rewarded.

In particular, role-players recounted stories of outbursts of anger occurring during the earliest years of working in the simulations (around 2008) during intense sectarian tensions in Iraq. I myself never witnessed any such overt moments of rage during the simulations I observed (2011–2013). In one such instance, at the height of the war in Iraq, I heard about how "Ahmed" had stared at an American soldier with fury, causing a spontaneous fistfight between them. Another role-player described the incident: "Ahmed decided to glare at the soldier for a long time. The soldier dropped his gun and jumped on him and said, 'Come get me, motherfucker!'" Rather than inscribing the moment within the plausible simulacral, a potentially heated moment in wartime itself, both parties were told by their employers to control themselves.

When these volatile affects had appeared in the simulations, military personnel tried to contain them in various ways. In one related example, "Basheer," a former Iraqi role-player had been promoted to the level of cultural advisor and occupied a well-respected roost and the ear of the military personnel. He had worked in the simulations at one of the military bases in 2008, immediately following the height of sectarian violence in Iraq, and tensions were particularly high among role-players, as well as between role-players and soldiers. Basheer noticed that during the mock demonstrations, the role-players actually appeared to be filling with rage:

> "They would shout, '*Barah, Barah, Amrīka*' [Get out, Get out, America], and they were so angry. They were angry for real."

> "How did you react?"

> "Of course I did not report them. I just told the military, you should take that line from the script."

In another such instance, there was a conflict between the company that hired them (many of the employees of which were ex-military) and the role-players. Basheer explained:

> "One day, the role-players overheard the site manager say a terrible thing. He said: 'These fucking Iraqis. I killed Iraqi children. Iraqis are animals.'"

> "And what happened?"

> "After this story spread, 250 role players stopped working: they were boiling."

The program manager at the company approached Basheer and asked for help, proposing that the role-players come up in a line to the site manager and receive his apologies, asking Basheer to convince the role-players. Basheer explained:

> "I started talking to the educated people. I said the work here isn't about that site manager. We weren't hurting him by not working. We were hurting the Army. This work was about helping both sides, Iraqis and Americans. It wasn't about losing money here, but about losing lives."

> "And how did other role-players react to what you said?"

"I got ten to twenty who agreed with me. We went out to talk to the 250 people. They agreed to continue working if that site manager got fired. He didn't get fired in the end. He got moved."

In this instance, the warranted fury of the role-players was diffused: they wanted to keep their jobs, but more potently perhaps, they were assimilated within military logics. During the height of the war in particular, although there was anger among the role-players about U.S. occupation, they were asked to sublimate it to help "both sides." There was no room, thus, for pure anger: not only against the single individual who had perpetrated the act, but against the Army and its affiliates as well. Role-players were required to position the contractor as a single bad apple, rather than expressing fury over the American military's occupation of their country. As he mulled over the incident, Basheer tried to reckon with and contextualize the contractor's behavior, even becoming in moments, an apologist for him:

"Another time, this site manager was talking with a captain, and the captain was crouching next to him, and he called the captain an animal. It made me think maybe he just used this word lightly."

Indeed, although anger and violence were essential within a number of the role-plays I observed, they had to be enacted in predictable and contained ways and could not exceed their boundaries and become "real." In recent years, professionalism has been emphasized and outbursts have become extremely infrequent; role-players fear losing their jobs and thus hew very carefully to military parameters. Meanwhile, affects of intense grief were applauded. Like military quests for and distancing from epistemological realism, the alternate courting and controlling of affective realism produced ambiguities and contradictions in the training.

·   ·   ·   ·   ·

One evening, a group of training soldiers failed an exercise and were punished. They had not gone the long way around and taken the right precautions, but had instead gone a more exposed and less safe route. Consequently, insurgents had (notionally) killed a number of the soldiers along the path. In order to make a pedagogical point that would be remembered, the remaining soldiers had to measure and dig twelve mock graves for those who were killed. The notionally dead were asked to lay shirtless by the fire and then to compose their own eulogies.

I sat with the Iraqi role-players, and together we watched this succession of events. It was twilight, and the light was mellowing. We watched the soldiers hack at the dirt and soon they were digging from within the grave, making empty boxes in the ground. The upturning earth of the woods was clayed veins of orange. They began to sweat and crack jokes: "Dude, you're a big dude! You're never going to fit in there!" I circled the graves and took photographs with my iPhone, stunned and unmoored by the spectacle. As the light waned, the soldiers who had been "killed" were told to compose their eulogies with a partner. In a surreal hum, I watched the speaking bodies, glowing by the fire, exchange details about themselves that might go in a eulogy. I listened to fragments amidst the hum: "I'm firstborn. I've got two sisters," said one. "I like sushi," offered another. It was getting darker out, and the role-players began to assemble their dinner adjacent to this exercise, which had gone on longer than expected, due to the punishment. Many of their wives had packed food: there were Tupperware of *dūlmā*, the grape-leaves and hollowed onions and eggplants filled with lemon and tamarind-paste-infused rice and meat; and biryani, with chicken and rice. I was eating with my role-player friends but meanwhile staring, riveted, and with obvious malaise at the staged deaths at the soldiers.

"Relax, Nomi," said one of the role-players.

"Isn't this frightening to watch, isn't this strange? What is it like for you to watch these fake stagings of death after having been in a real war-zone?" I asked.

Most of the role-players in this group were in their thirties on so. Many had worked as interpreters for the U.S. military in Iraq, but had also experienced U.S. sanctions and witnessed the Iran-Iraq War of the 1980s.

"Come on. Nothing about death is strange. My cousin died in front of me. We couldn't bury him right away, we had to put him on ice."

.   .   .   .   .

The simulations and exercises frequently brought role-players in contact with notional mourning, injury, or death—either witnessing it, like in the fake grave punishment, or theatricalizing it. In the latter instance, the Crying Room was only one iteration. They also participated in "Mass Casualty Events," and howled in the aftermath of the many deaths simulated around them and in some instances were adorned with graphic fake wounds. At some bases, the role-players wore laser-tag-like belts (Multiple Integrated Laser Engagement System or MILES) that illuminated to register their own notional deaths. As role-players with varying pasts in Iraq performed these representations of extremis, and as they negotiated their employer's Left and Right limits, they performed and/or experienced a complex range of affects. I frequently asked role-players if they were emotionally affected by enacting their own deaths so recently after war. The typical reply invoked a habituation to death due to Iraq's history: "No—we are Iraqi. We are used to it." One role-player, "Hussein," answered: "No, of course not. You're surrounded by death." This habituation to death and the repetition of weeping on a military contract suggests that the role-players might be conceptualized here, in part, as something like salaried lamenters. Like professional mourners, their responses suggest that they typically do not, or do not necessarily, internalize their own affective performances. Typically as the role-players wailed and hit their thighs and heads, the resulting scene was eerily stylized, insinuating the presence of an emotional buffer between the actors and action. However, in some instances, the summoning of affect for their performances engendered a more potent experience for the role-players.

A number of the role-players interviewed had extensive experience with war and expressed moments of intense emotional resonance. When a role-player embodies a role with convincing intensity, the other role-players call this 'aīsh-bi-dūr ("living-in-the-role"). One role-player, "Osama" explained: "It's not acting, man! It's remembering. It's hard to make your tears come out." Mohammad, another role-player explained: "There were moments in the simulations when there is a tickle. It tickles you back into Iraq. It would become real." He described his experience working in the Crying Room: "I was playing the brother in the room with the crying women. It was emotional for me. Real tears came into my eyes. I have seen so many women crying like that [in real life]. I know people who have lost their kids." Mohammad explained that it only took a short time to enter into the emotional situation of the scenario:

> "It takes me only three to four seconds for it to be real, to go into myself and act like it's real. Once I was role-playing a doctor. I was actually crying. In the scenario, there were sick kids and women were dying. I was the doctor, and I had no medicine at all. I put myself in the situation."

> "And how do you handle that line—between enacting a role convincingly and being overtaken by it?"

> "When I go too far in, my brain tells me to stop. I tell my brain to stop. It is like listening to a song, and it was your song with your girl, and you broke up. And then you listen to it, and you feel it. That's what it's like, sometimes."

Occasionally, role-players who had endured the most acute war trauma, such as death threats, kidnappings, and near death experiences, noted that simulations could be emotionally perilous. In these instances, a role-player is momentarily overtaken by the potency of the role, its echo of the wartime lived—what I call, in an inversion of the expression role-players used, being "lived-by-the-role." In this instance, the role-player relives the particular precarity of their wartime position, accused from all sides.[1] Within the military space, power relations render different human beings differentially grievable and some seem to be perceived as more expendable than others.

For example, "Sumayya," a role-player who had had traumatic experiences in Iraq when she was an interpreter was "lived-by-the-role." She had been role-playing an interpreter all day and was told that there was only enough protective gear for the American soldiers. She told me afterwards that this military choice had sent a shock through her. How could the soldiers bring her into war with them without protecting her? Why wasn't her life as valuable as theirs? Indeed, why, essentially, we might ask, was her life being imagined by the military trainers as "ungrievable"? To render a person ungrievable is to potentially produce a psychic wound producing melancholia (Butler 2009).[2] To continuously theatricalize this expendability in a role-play is potentially a reproduction of that wound.

The following day in a simulation, Sumayya encountered a role-player wearing a *kūfiyyah* scarf over his eyes, a costume she associated with insurgents in Iraq, and this upset her. She then spent several of the evenings after the role-plays distressed, away from the group. I asked "Hussein," another role-player, what he thought had happened to her and he replied: "*She was living it.* Her reaction then would be the same as now. She was shaking."

A few role-players told me in various terms about the dangers of falling into the role and remembering memories from wartime. One role-player, "Billel," described the need to engage in *riyāḍah al-nafs*, an athletics or pedagogical exercises to buttress the self during a simulated strike. Billel described this as trying to ward off something frightening—like a jinn or a ghost, to avoid falling into a "black room."[3]

> "Sometimes I do this in the war game. It is like this: when there is a strike, you open the door, and the room is black shadows. Now: terror. The next time, you open the door again and nothing is there. You cross a second distance. You say to yourself, I am in control of myself."

Another role-player, "Ahmed," also described the peril of being seized by the role and reentering an excruciating form of wartime precarity. He described a dream where he was working as a role-player in the simulations and fighting a black dog:

> "I was at [the exercises]—it was a normal [exercise], except there was a giant black dog behind me, and all the others ran away, and I was all alone, and I had my gun, and I was hitting the dog, but there was no effect. And he was hitting me. I continued to hit the dog, and there was no effect. And then I woke up."

> "And does the black dog relate to your own life?" I asked.

> "I have no black dog here in America. Many Iraqis here do. Many of the Iraqis I work with at the military base. They have an internal struggle, like two people fighting inside of them. They feel like the Americans destroyed their country, they did bad things in his country."

> "Do you feel differently?"

> "I don't have this feeling. Some people just want to get their citizenship and leave. I don't feel this way. I don't always think properly though, because of the past. I need to wash my head."

Ahmed ascribes the black dog to a condition within *others;* nonetheless, he still thrashes the metaphorical figure in his dream. He is unable to make headway, but he refuses to give up, acknowledging the complex encroachment of the past into his current life: "I need to wash my head." Loss and vulnerability persist, as many role-players negotiate deep wounds and in some instances internal contradictions—as well as the aftermath of a deeply precarious in-between position where they were required to carefully perform their affiliations in different contexts in order to survive. Alternately maligned and accused of being spies by both the Iraqis and the Americans for whom they sacrificed their safety, their lives were treated as particularly expendable.

Although these moments of witnessing someone being lived-by-the-role occurred occasionally in my two years of fieldwork, role-players more typically dealt with the simulations in other ways. Rather than tumbling into entry-points of verisimilitude to their wounding experiences in Iraq, they more often lightly described the simulations as tiring, something just to get through, or generally responded with affects of indifference or laughter. As Billel once told me, as he explained his strategies of self-control in the simulation, "these people (Iraqi role-players) are turned inside-out. At the same time, they can laugh and cry. They live both states, both stages. In the past, things were harsh. Now, you must get happy over anything. You must laugh."

During the simulations, role-players must act out Middle Eastern villagers "authentically" not by their own measures, but rather within prescribed military terms. The simulations are spaces of immense and uncanny repetition and artifice, provoking sensations of familiarity alongside undecidability.[4] The uncanny sensation is a catalyst not only for malaise, but also in some instances for laughter. The simulations are littered with objects and representations that provoke such sensations of uncanniness: from partial or prosthetic body parts (wound kits with disembodied pancake flaps of skin decorated with blood and gangrene) to the reanimation of the "dead." For example, a laser (MILES) belt that indicates whether someone is alive or dead can sometimes go haywire: people can be reset and brought back to life. Meanwhile, as military architects work to create an "authentic" Middle East, they often mire the simulations in artifice, creating comedic and sometimes uncanny disjunctures for role-players. Rather than a simple surface "pastiche" of the Middle East, in Jameson's sense (1991), the *copy* reads as a parody to those who have been "there."[5] Indeed, the settings are often contrived-looking stage-sets, littered with objects that index their own artifice. For example, at one fake village, there was a chicken coop with a decapitated Styrofoam chicken—a joke and a register of not quite alive, not quite dead—a reminder of the not quite real, but not totally fake. As they are asked to represent their culture synchronically, pruned of any relationship to the outside world; as they are turned into archetypes inside an archetypal village; and as they act out death and precarity so many times that these losses and injuries turn even more into estranging archetypes, they laugh.

The seemingly recognizable and "real" (the referent of the Middle East, of war, of death) continuously becomes laughably fantastic. Laughter acts as a response to the gap between alive and dead; moving and frozen; the hyper-artifice of "Iraq" and the painful "real" of an Iraq to which these Iraqis cannot at present return. As they repetitively embody market-goers, prisoners, beaten wives, mourners, and militants for the training soldiers, the ossified postures, characters, and interactions they stage sometimes trigger laughter—thereby interrupting those postures. Henri Bergson proposes that a certain kind of laughter is provoked by "mechanical inelasticity, just where one would expect the wide-awake adaptability and the live pliableness of a human being" ([1900] 1914, 10). The repetition of the face seals it into a machine or death mask. Ironically, however, such archetypal faces produce laughter, interrupting their stasis: "Some faces seem to be always engaged in weeping, others in laughing or whistling, others, again, in eternally blowing an imaginary trumpet, and these are the most comic faces of all" (25). To this end, laughter might act not only as a kind of reprieve, but also as a definitive crack in that mask. Most crucially, laughter decisively resituates the role-plays as artifice, enabling the role-player to create their own real amidst that artifice.

The machine thus turns out to be made of flesh. Role-players inject new ways of being, in part through laughter, into their performances. Those interjections show the limits to a military fantasy that human beings can be wholly resourced and turned into technologies. Still, they may have little impact on the military's training structure itself.

. . . . .

There is a crescendo of sobs and Arabic in the Crying Room. Each team of soldiers reacts differently to the cacophony, their captain maintaining calm or becoming flushed and anxious. For the soldiers to succeed in this simulation, they must first establish rapport with the mourning women, then notice that the house might be connected to the town militia and procure intelligence. Meanwhile, the hubbub rises in the simulation: sometimes one of the women runs up to a soldier and entreaties him furiously in her own dialect of Arabic, the specific content of which he typically cannot understand.[6] Amidst the clamor, the captain tries to build a relationship with the women.

But the role-players have been screaming all day. Amidst the ennui of the endless mechanized repetitions, hoarse throats, and the gap between this coffin and the coffins they saw in Iraq, the role-players sometimes stop wailing the classic words of grief. In such moments, muffled laughter can commingle with their cries as a role-player conceals her faces with her veils. In one instance, one role-player shouted in Arabic: "How long until lunch?" "Lubna" told me: "Say anything as long as it's in Arabic! Once I yelled: 'I want a hamburger,' forgetting the soldiers would recognize the word, and I clapped a hand over my mouth." Similarly, in the midst of a Mass Casualty Event, another role-player, "Salma," jokingly yelled out at the "perpetrator," "That's my uncle!" This precipitated suppressed laughter amidst the screaming.

Indeed, the work shifts are long and taxing, as the role-players act out war wounds, their mouths frozen into Os of grief, their voices literally nearly gone from crying. But wanting the shift to be over and wanting a hamburger and *saying it* is an injection of the living, actual body as a retort to the mechanical body. Making a dark joke about one's uncle, amidst a narrative where all Iraqis are potential adversaries, is also a refusal: the role-play restitches itself into a moment of rebuke.

Satire opens up the possibility here for shifted moments of subjectivity and potential political comment. However, such subversive moments do not ultimately overthrow regimes of power and their representational imperatives. Such moments like this (which I observed or which were described to me) were intrinsically very brief and self-correcting: role-players did not want to lose their jobs, and interrupting the flow of the simulation constituted a threat to that work. After the hamburger incident, Lubna quickly covered her face with her veil. When Salma yelled in Arabic that her uncle was a terrorist, I looked at her wide-eyed and she and I both suppressed a laugh, then the simulation continued unchanged. Likewise, when role-players experienced epistemological and representational surfeit, they often kept it to themselves. I did not observe any circumstances where either training soldiers seemed to understand the brief Arabic retorts between role-players, or that military personnel stopped the role-play in reaction to the brief overflow of affect. The soldiers did not seem to notice such momentary disjunctures, or they ignored them; they were quickly shuffled onward to the next simulation.

These wartime intermediaries are being asked to enact, without rupture or divergence, a military fantasy of a range of readable and useable Middle Eastern individuals in wartime. Each character encompasses a recognizable type and a lesson that might be used by the military. In the Crying Room, for example, there are "mourning mothers" and "sisters" who might be willing to offer intelligence if that would make their villages and children safer. The soldiers must establish a relationship with them, evince pathos when they weep, and convince them to offer assistance to the U.S. mission. In military logics of counterinsurgency, all local individuals in wartime are "good guys," "bad guys," or "fence-sitters." Fence-sitters can be wooed over to the American cause and co-opted as "good Arabs" (Cohen 2010).[7] In this exercise, the women are potentially the "good Arabs": the soldiers' success depends on their co-optability. In this constellation, the women must be willing to appreciate the rightness and justness of the American cause, the appeal towards a universal good, posited as implicit in the U.S. occupation of the mock Middle Eastern country.

That archetype marginalizes the much more complex position and personal and familial stakes for the actual Middle Eastern individual in the middle of this geopolitical and ideological battle. In the scenario, the men of the house are affiliated with the militia. For the U.S. soldiers to be effective, they must procure social information and intelligence from the women. Yet for those women to share information is for them to risk everything and be accused of betrayal by their own families. Given their own wartime histories, many of the Iraqi role-players are intimately familiar with those stakes.

Perhaps at least in part due to that knowledge and the unusually complex and in-between subject positions they have lived, rather than remaining wholly compressed into a military archetype, something else happens. Lubna displaces the archetype of mourning-mother-good-Arab, and instead yells out that she is hungry. Salma, meant to act out an Iraqi woman potentially affiliated with the terrorists, jokes about her uncle. While a military regime is not overthrown, the moment perhaps contorts that regime, enabling a "subversive and parodic redeployment of power" (Butler 1990, 124).[8] These momentary eruptions where laughter and screaming commingle are caused first by the act of embodying that archetypal face—but they also *interrupt* it. Just when machine-like repetition seems to refute the variability of the human being, "living life" itself returns in full force. Through laugh-infused wailing, role-players find momentary means to comment on that artifice, creating a rebuke both to being turned into automated archetypes and to inhabiting wartime wounds. The outburst is something akin to Donna Goldstein's (2003) "laughter out of place," where humor forms a retort to suffering and precarity. Through laughter, role-players insist on a changed form of being alive after war's injuries, effectively refusing to robotically enact the perfectly co-optable intermediary. One role-player, "Hussein" coined a name for this subversive affect. He described how role-players had reacted during a simulated funeral: "They brought him on the stretcher. Some people were screaming and moaning. Others were *laughscreaming*."

In this moment of affective surplus, the role-players distance themselves from embodying military archetypes and becoming slotted into the violent fixity of military categories. The person knows or feels more than the military narrative of their experience can accommodate, exceeding the constricted functions prescribed for a hired cultural tool. Additionally, the laughscream acts as a refusal to be lived-by-the-role and the role-players' fraught wartime pasts. For those accused of betrayal and marginalized by both Iraqis and Americans, that past is terribly painful. As "Bilal" explained, reflecting on the harshness many Iraqis had endured: "We are turned inside-out. At the same time, we can laugh and cry." Indeed, for Iraqis who worked with the U.S. military, it has been prohibitively dangerous to return to their former home, particularly amidst the ascendance of the Islamic State in 2014. Meanwhile, due to their wartime choices, many negotiate ongoing ambivalence and feel stranded between nations: although they were frequently ejected to the peripheries of their countries for working with the Americans, many strongly identify with Iraq and are ill at ease with full assimilation in America. As they continue to work for the U.S. military, some conceal that work from their families in Iraq, grappling with how they might be perceived. Amidst these tensions, the laugh-scream functions as a charged and ambivalent intensity—something like a brief *fuck-off* to the absurdities of the system that uses them but does not value their personhood.

Laughter rises to confirm that, for the role-players at least, "Iraq" is not Iraq.[9] As fake guns sound, role-players repeat themselves, becoming increasingly estranged from the original object. Yet, through laughter, the archetypal and mechanical face of country and person give way, to Iraqis who live impossibly hybrid and ambivalent lives in the America to which they at such great cost, aligned. By "parodically" redeploying power, the mechanical performance of death becomes a complexly subversive act, which may momentarily reinsinuate life for the role-player. In (quasi-) inhabiting death without dying, in staging death amidst a life of precarity, the individual perhaps experiences something like a Bataillian jolt of laughter.[10] But whereas for Bataille, the laughter is produced by an experience of senselessness or meaninglessness, for the role-player, laughter makes its own sense, contorting military representations.[11] As the subject acts out death, and as the sense of the military law and its archetypes momentarily disintegrate, laughter becomes a principle of at least momentary animacy. Laughter here—and the risk of allowing it even briefly outside the proper time and place—is a jolt in the body that acts something like a resurrection from the death of death and an undoing of the death produced by military reification.

Theatricalizing death produces various surfeits for role-players, from grief to hilarity—as moments of apparent indifference and boredom fluctuate with experiences of unbearable pain and subversive laughter. In recent decades, social theorists have shown affect's role in subjectivity, action, and agency, emphasizing the "autonomy" of the body in its interface with the world (Massumi 1996; Stewart 2005), and the role of affect in making social projects effective (Mazzarella 2009), even proposing that affective forces might be "capable of overthrowing" discursive regimes, initiating, if only momentarily, new modes of living in the world (Guattari 1995, 19).[12]

The role-players' laughter *matters*—perhaps it can even be described as a kind of self-preservation of their subjectivity or release amidst fatigue within this context of labor—but what does this affective moment mean in the context of the military project? Laughter and joking are hardly among the prescribed functions of a cultural training tool of mourning. However, even as the role-players experience the incongruities of how the Middle East is represented, or even as they laugh, arguably little changes for the *user* of that tool, the training soldiers, who are quickly shuffled onward to the next simulation. Arguably, the military structure is neither undermined nor reinforced through the affect of those who work within and for it: rather, it continues seemingly unchanged. Military systems and notions of the world continue undisturbed.

While acknowledging the potential agentive importance of such moments for the role-players themselves, there may be little ultimate impact on the military structure itself—or for training soldiers, a step removed from these moments of surplus the role-players are experiencing. In this instance, affective maneuvers recuperate subjectivity; they don't modify structures, a reminder to locate affect within power relations, inequalities, and violence (Povinelli 2011; Biehl 2013; Adams 2013; Berg and Ramos-Zayas 2015). After a genealogy of thought that has examined what "structures feelings" (R. Williams 1958) and then queried "what feelings structure" (Richard and Rudnychyj 2009), we land elsewhere: affects may offer an altered possibility for those experiencing them, but may "structure" very little within the military training itself. Theoretically and on an existential level, such affective interruptions and overflows throw into question the contradictory assemblage of the "human technology," and the extent to which human beings can be resourced in this manner. However, ultimately these individuals, harnessed as cultural tools, largely remain working within the terms forged for them, as soldiers continue to rehearse the operationalization of culture.

"Bill," a retired American soldier involved in the trainings both in advisory capacities and as a role-player himself, explained: "Unless something interrupts the goals of the training, then whatever is done by the role-players, who gives a shit?" In his opinion, it seemed unlikely that the training soldiers even noticed such moments of disjuncture: "Think about where their mind is at. You think they're going to catch those little nuances? They could care less. Those things [the Iraqis saying something extra or laughing for a minute] have no value." "John," another soldier involved in crafting and supervising trainings, agreed that moments of disjuncture for role-players did not interrupt the training, but proposed that such instances might "in some cases reinforce training objectives." Moments of puzzlement might cause the training soldier to "wonder if he had redirected the cultural exchange in an uncustomary way." However, he too concluded that: "If a tree falls in the woods and no one is around to hear it, does it make a sound?"

I asked John about the possible relevance of role-players' internal experiences in the role-play. I described the periodic jokes within the simulation, noting that they were particularly ironic and complicated if one knew the role-players' backgrounds. Some struggled with working for the U.S. military, given their mixed feelings about American interventions in the Middle East. Some even hid the work from their families back home. John found my observations tangential: "If there's trauma behind it or it's just sass, who cares? What's the 'so-what'?' It's surreal, I'll give you that, but that may be all." Bill explained: "All that matters in the training is the soldiers get to practice rapport. All the rest is background noise."

For "John," these moments were potentially akin to little more than background noise, if they were noticeable at all. Indeed, even if the soldiers noticed that something was a little "off," more crucially, they were not equipped with the context necessary to read such a moment of disjuncture. Soldiers were dislocated wholly from the complex motivations and internal worlds of the role-players employed in their service. Ultimately a role-player's experience in the Crying Room or in a Mass Casualty Event was only truly legible to someone who could contextualize that particular individual's politics and potentially deeply complex relationship to politics and country. Was a role-player simply bored by the endless repetition of the simulation, or instead negotiating an anguished internal contradiction about the American presence in the Middle East? A lack of adequate Arabic to perceive such a moment was only the beginning of the story; rather, layers of depth and complexity were occluded from the training soldier. Ironically, a cultured body hired to help soldiers become "insiders" instead remained an inaccessible surface to those trying to go *in*. In the Cultural Turn, the subjectivity of its subjects is irrelevant.

Nonetheless, the training mechanism continues to function, preparing the soldiers for war, largely without disrupting their notions about difference in the world. The soldiers enter the Crying Room, perform empathy for the mourning women, extract information with varying degrees of success, and walk into the back room and find the weapons, confirming the women are exactly who they suspected all along. Re-set. In the training machines of the Cultural Turn, both the Middle Eastern role-players *and* the soldiers are tools, part of a military system that resources human beings in the service of Empire.

Still, a last ironic kink persists. Military cultural trainings, at bottom, seek to prepare the soldier "for the real thing." Yet, the simulations embody a charged irony, where those who are militarized as human technologies, in part, to be-themselves, know that they are anything but. This gap suggests that the military edifice itself is perhaps out-of-order, even on its own terms. Namely, a training mechanism to render the human terrain legible and containable, which effaces its own moments of excess, perhaps encounters its own limits when the soldiers arrive in the warzone. As one military interlocutor, who problematized the fixed categories of some of the cultural trainings, cautioned, by reading me this line from a William Meredith poem: *"Things are not orderly here, no matter what they say."* The laughscream is the hiccup in the machinery that may be imperceptible at the time of the training, but that may signal the structure's potential breakability in the war beyond. Likewise, this hiccup foreshadows the military objectification of the human beings the soldier encounters in the warzone, and thereby the farce of the Cultural Turn at large—a strategy that emphasizes humaneness but erases the very individuals it purports to project, the so-called locals in the warzone.

·   ·   ·   ·   ·

The strange hiccup of the laughscream is an affect not confined to military simulations but is rather a mobile concept, a traveling charge. Partial kin to Turner's (1969) *communitas*, as new solidarities or in-between worlds might be forged, or to Bakhtin's (1984) *carnivalesque*, with its possibilities for satirical undermining of authority without unmaking the structure, and partial kin to William Mazzarella's revived notion of mana (2017), an energetic rush of energy or potentiality that is both "intimate" and "impersonal," the laughscream journeys between contexts and geographies. The laughscream might occur in isolated moments (such as in the military trainings, or in an effigy-burning in North Africa that I will describe shortly) or it might become a more diffuse affect that comes to characterize a time period, changing the temperature of a room in a grim political climate (such as during the Trump administration). The affect is frenetic and the playful (laugh) veers into a darker underbelly (scream).

I noticed a kindred affect amidst very different structural circumstances when I was doing fieldwork in a small Jewish community on the island of Djerba, Tunisia, as they burned Purim effigies of the villain Haman: a moment of commingled jubilance, anxiety and fury, the overcoming of historical persecution. In the military Crying Room, role-players enact their own subjugation (their country has been occupied) as well as their grief (a child from the family was run over by an American tank). In the Haman effigy, the ritual participants enact their own triumph (which shadows both ancestral memories of persecution and contemporary anxieties around being a minority in the Arab world). In each case, a population uneasily inhabited a zone where they felt, to varying degrees, to be subaltern. This state (subalterneity) contorted into a new affective form when pressed into an intensified and closed circuit with ritual and repetitive qualities (that is, the Crying Room, or the Haman effigy). The laughscream is the charged outcome, which does not alter structural circumstances of its participants.

Trump's America, for example, might be described as an extended laughscream for those resisting the politics of the day: there was in it both comedy and dread, a feeling that even as we speed-dial our representatives all day and even as we make a small gain, other inhumane polices are, at the same moment, being put into place. In one example of innumerable, Trump's family separation policy has led to undocumented two-year-olds being required to represent themselves in court. In a John Oliver segment on the topic, the audience's uneasy then almost relieved laughter offer something like a moment of rebuke, a change in the pressure in the room. But make no mistake, its underbelly is screaming, if not mourning.[13]

Dominic Boyer and Alexei Yurchak point to the possibility of satire in generating an "alternative aesthetics and practice of political critique" (2010, 213). Without overthrowing a political order, such moments of laughter might do "real political work, fostering the development of new subjectivities" (Bernal 2013, 306). In the darker affect of the laughscream—where comedy commingles with horror and uneasy mirth is fused with mourning—a room might temporarily reset. However, the conditions that room is subject to: they crank on.

[Field Poem Fragment]

The story says we are in the country of Pineland: grassy roads
    curving in, named
for longleaf pine, loblolly pine. Sassafras, blackgum, slashpine,
clethra sharp as pepper, shallowing
the land's breath.

[Field Poem Fragment]

Sense   is an edge, see
        if you dare look
        over into the white
falling

[Field Poem Fragment]

The men make a circle
The pines make a circle

6   Becoming Human Technology

We are in the woods of Pineland, where the grasses seethe with ticks, and I am staggering with all my belongings: tent, water, and provisions, on my back. Once I cross into the guerilla camp, I must assume the role of a woman named "Gypsy" who wants to join the guerilla fighters. A military escort brings me part of the way through a path in the woods, carrying my rolled-up sleeping pad. He hands it to me, and gestures towards the rest of the path and a clearing ahead: "You'll find them there." I've been given a very brief sketch of the character I am to play, and I do not know what to expect, or how the training soldiers will receive me—a woman with her arms awkwardly full, suddenly emerging towards their many, in the woods. As I approach, I see tents scattered through the woods, a rustically built kitchen, a medic's station, a Pineland flag, a central firepit and clusters of figures. The forty guerilla fighters (all contracted role-players) and the team of a dozen or so Green Berets, who are training them to fight, are all men apart from the cook.[1] When I walk through the path and into the clearing, the men look up with surprise.

I am entering Robin Sage, the culminating two-week field exercise in Unconventional Warfare for the training Green Berets. It is the only time they undertake a similar exercise of such complexity, intensity, and length.[2] Robin Sage is conducted approximately six to eight times annually (with more than one hundred students each time) at the end of the Special Forces Qualifying Course. During Robin Sage, central North Carolina (spanning nineteen North Carolina counties and six congressional districts; Woytowich 2016) becomes the Republic of Pineland. Because of the integration into local towns, those who are role-playing wear specific costumes (hats and armbands), and all local law enforcement is briefed in advance.

Through the course of the exercise, they will train proxy soldiers (role-players) to undertake U.S. policy aims by overthrowing their own government. The woods are at the heart of the exercise, but it extends outwards: North Carolina is studded with guerilla base camps and camps for the countermilitia in the depths of the forest. Surrounding local farmland loaned to the Army is used for supply-drop zones in the simulation. Many of the adjacent towns, too, are involved: little bait-and-tackle shops volunteer to be meeting grounds for the resistance; greasy spoon restaurants accept *don*, the Pineland currency, for biscuits and gravy (and are then compensated by the military).

These particular farmlands, woods, and towns have been home to Robin Sage since 1974.[3] Robby, a ruddy and bearded man who described himself as one of the first role-players, told me that his town was hardly separate from Pineland.[4] He walked me around the town to show me the spots that evoked memories. On the white cinderblock wall of the grocery store, the Army would project PSY-OP videos on the side of the building, and the town would watch in a dirt parking lot. His town had a two-cell jail that had never been used before; during training exercises, soldiers would catch adversaries, "hog-tie them on the sidewalk" and imprison them there.

Robby's first and most personal Pineland memory was from when he was small, before he had even started school:

> "It all started one Sunday morning. Mom came in and woke us up and said we're getting ready to go to church. And she said there are some soldiers out there in the yard, alright? And so we went on outside and stood on the porch and there were jeeps around and a tank pulled up."

> "And what happened? Did you get to stay home and watch?"

"So, me and my brother ran back in the house and says, 'Can we not go to church today, we wanna stay here!' And she says, 'Yea.'"

"And then what happened?"

"Anyhow, we went outside and they were setting up in their battle positions, and I don't even know this guy's rank or nothing, I just remember, roughly, this guy's facial features. He pulls up in a jeep in the driveway, and he says, 'Would you fellows like some candy or something?' And he made both of us a C-rat can full of hot chocolate. And we're sitting there eating our candy and drinking our hot chocolate and we're just taking it in, and by that time, they had started their battle . . . "

By the time Robby was ten, he began to intrude within and even tinker with the outcome of the training scenarios. He dressed himself in his father's National Guard uniform and made himself a red armband so he could be a guerilla fighter, or "Little G." He described how local children would derail the opposition force and play jokes on the soldiers:

"We were armed! 'Cause, you light up a ton of firecrackers in a side ditch, it sounds just like an ambush going off. And when you have three or four guys just doing it at the same time, it sounds pretty good."

"Then what would happen?"

"So we would wait for the enemy forces to drive by and we would light them daggone fire crackers and throw them in the ditch and it would be just like them going through an ambush, and they would stop what they were doing, and they would chase us, but we would just take off."

In the dizzying pivot between the real and the simulated, it is hard to know who might be a guerilla, who is role-playing a guerilla, and who is a child faking it. When driving inside the depths of Pineland, one's GPS does not work, and it is easy (or, at least, it was for this anthropologist) to circle around interminably. A major told me how soldiers would sometimes "fall into" Pineland: "There would be guys pulling security with a bunch of shrubs, or they would think ruck-sacks were dudes. One guy was talking to a tree; he said, No Sergeant, Yes Sergeant." And role-players would some-times gesture at the wall of pines, joking: "There's no way out of here!" When you enter these grounds, the pines seem to contract around you.

·   ·   ·   ·   ·

Before they enter Pineland, the training Green Berets are reminded that their mission is "Left of Bang," creating the silent infrastructure for the big war. Their instructor tells them: "No one knows you're going to Pineland. We're not putting a flag up. There's no magic hour in this situation. Figure out the ground truth. You have signed up for Unconventional Warfare." As a lieutenant-colonel told a new class of trainees, as narrated in Dick Couch's *A Chosen Soldier:* "This is where you take everything you've learned and put it all together in an operational environment. We expect you to work with indigenous forces in an unconventional warfare scenario. . . . This phase is all about managing the human terrain" (2008, 288).

During Robin Sage, the trainees are assembled into twelve-men Operational Detachment Alphas (ODAs) or A-Teams, to practice covertly mobilizing a local band of guerillas to overthrow their own government. Once the exercise begins, soldiers are required to identify and assess the four main guerilla leaders in Pineland to see who is the best fit to be armed by the United States and offer recommendations to their higher-ups. A military instructor explained: "In unconventional warfare you use force multipliers through the locals. We send out twelve guys rather than a brigade. Ultimately, you're there to fight but you're not fighting. You've persuaded, convinced, muscled people to do that for you."

This mission requires that the Green Berets mute their presence both geopolitically and culturally. One morning, I stood outside with a group of military personnel who coordinated the exercise. They were smoking cigarettes and explaining what made Special Forces *special* compared to a conventional soldier. One described the deployment of a Green Beret revealingly as linked to a capacity for profound entry into the local culture. He explained: "We become them." On another occasion, a major boasted: "Our team members become honorary members of the shuras." An attention to local mores was understood as essential in this duality of interiority and muted presence. Another major elaborated:

> "We understand culture, and we try to adhere to belief systems. It's less invasive, and the people are more accepting of us. In what we do, the Host Nation forces are the lead element that people see. There are Americans too, but we've got beards, and we look as much like them as we can."

> "And why do you grow beards?"

"Growing beards diffuses the issue. It is not a matter of *hiding*. For example, if Arabs wear *kūffiyahs*, it might bring out xenophobia for Americans. But if they wear a baseball cap, it diffuses it."

"And what's the effect of a Green Beret wearing a *dishdāsha?*"

"If we wear a *dishdāsha* to a meeting, it goes more smoothly. When we wear it, it's the equivalent of wearing a tweed polyester suit with a mismatched tie. They look at us and think: He is trying. The dude is a little off. But he's still trying."

He then extended this notion of muting the impress of his presence in the warzone into a more startling metaphor, explaining the point of the performance:

"We are sharks wearing skin-suits. We assimilate to kill whoever we need to kill. Sometimes we use people. We never lose sight. We swim through that sea. We zip up our skin suit. We sharpen our teeth. We wear our *dishdāsha* to get close enough to stab them in the back. If I have to become a chameleon to get to them, watch me."

.   .   .   .   .

Deleuze and Guattari describe a "Binder God" as one of the poles of the state's power, who "emit[s] from [his] single eye signs that capture, tie knots at a distance" (1987, 424). When the Binder God sends men to war, no weapons are required. Rather, his eye encasts that which it gazes upon, sending all into a "petrified catatonia" (425). In the theory, the state seeks to encast and interrupt or "striate" all moving, "smooth," and "nomadic" flows under its purview, from persons to commodities to capital. Through striation, the state seeks to regulate migratory bodies as well as nomadic thoughts—that is, the uncategorizable and dissonant musics of the mind, ideally even capturing the human interior itself.

The "nomadic war-machine" begins as human flight from the state, and only becomes war itself when it collides with and is appropriated by the state. At this juncture, the war-machine becomes a state-driven machine. In the case of the Green Berets, after capturing that machine, he continues to mimic the nomad; he is still a shark, but he zips up his skin-suit and makes his way into the war.[5] As I will describe, both the conventional forces and to a greater degree, the Special Forces, have sought to stealthily co-opt the nomadic war machine.

In his writings on Iraq and adopting lessons from prior counterinsurgencies, Lieutenant-Colonel John Nagl asked that the American military's conventional forces learn how to "eat soup with a knife," adapting quickly to circumstances and learning on their feet (2002, 223). The *Counterinsurgency Field Manual* of the U.S. Army Marine Corps, updated in 2006 and then again in 2014, reads in part like a doctrine of mirrors: they adapt, so we adapt—faster. "The speed with which leaders adapt the organization must outpace insurgents' efforts to identify and exploit weaknesses or develop countermeasures" (Department of the Army 2006, 7–46).

In Unconventional Warfare, the Green Beret takes these logics of agility and diminished imprint further, called upon to become a "chameleon, to get to [the locals]" as the major described. Through these labors, the soldier seeks to mobilize striations that camouflage themselves as smooth space, by instrumentalizing social relationships. The Green Berets, who often describe themselves as rebels and cowboys, a space apart from the conventional forces, may be read as nomads seeking to erase their geospatial and cultural striations, to "become imperceptible." As they seek to become insiders within geopolitical and cultural space, they labor to gain ever more intimate entry, eventually into the intentions of their local interlocutors, in order to predict and maneuver their action.

·   ·   ·   ·   ·

Pineland is sometimes called the "Monkey Bars of the Special Forces"—the place that makes the Green Beret. The fictional country is malleable and in some ways generic: allowing for different cultures and different skills to be taught, depending on need. And Pineland has room, a military thinker involved in the predeployment training scenarios told me, "for anything you could want, every crazy thing that is happening in the real world. In the beginning, the scenarios were more like the Cold War. They change as the conflicts change. Now they are more about the Middle East and irregular war." Another theorist involved in the scenarios explained: "The enemy now is operating in his own country in his own population, and he has much greater sense of what's going on than we do." Accordingly, in the last ten years, military predeployment trainers have shifted the makeup of the continent of "Atlantica," which is comprised of four countries (the United Provinces of Atlantica, the Republic of Appalachia, the Republic of Columbus, and the Republic of Pineland):

> "Now, there is a clash of many cultures on Atlantica. We wanted to create that nuanced, complex, hybrid kind of culture. We developed a number of insurgencies that were very unlike each other—insurgencies that were Islamic in nature, those that were racially oriented, those oriented on urban gangs, those focused on socialist policies."[6]

> "How did you see this landscape of insurgency compared to what came before in war?"

> "We tried to create insurgencies that were not about the typical Cold War bifurcation. Instead there were many, many different ones."

The post 9-11 military focus on irregular adversaries and insurgents able to blend into the population, and the resulting Cultural Turn, generated a world where embodied cultural knowledge became a theatricalizable commodity. Alongside the shifts in the conventional forces, Pineland's previously more fluid world of mock villages and role-playing was refigured into an enlarged political economy of contracts, which encompassed a range of actors, from retired Special Forces to Middle Eastern nationals.

The war game scenario of Pineland generates conditions for the practice of classic Unconventional Warfare as well as incorporating shifting geopolitical concerns. In the story, the United Provinces of Atlantica, invade the disputed territory of North Pineland Province (NPP) of the Republic of Pineland (ROP), a U.S. ally. The U.S. Special Forces infiltrate the NPP to handpick and then train guerillas that the Americans will work "with and through" to topple the NPP and restore the ROP.[7] As a further note, the NPP has many nonethnic Pinelanders in it (African and Iraqi immigrants), hence the presence of Iraqi role-players in the fictional country.

To restore regional order, the Green Beret trainees must read, court, and prime members of the resistance, assessing potential leadership within the guerillas to lead the effort. Later, trainees partner with the resistance leadership in the contested territory in order to provide training and military advisorship during insurgent combat operations, throughout which guerrilla groups composed of native and/or ethnic Pinelanders take the lead. Throughout the course of the training, the soldiers must secure their wartime intermediaries and harness them for American operational ends.

.   .   .   .   .

Ideal human technology within American military fantasy enables the U.S. soldier to become a geopolitical and cultural insider within the warzone, co-opting the nomadic war-machine. To this end, the wartime intermediary would: (1) aid the soldier in accomplishing American policy goals without interjecting back agendas that divert American goals; (2) bear the bodily risk of the venture; (3) render the soldier an insider with regard to the local culture; and (4) to both have a stable identity and to stabilize the identities of others, and thereby dramatically minimize the soldier's imprint in the warzone.

A major assisting with the trainings, "John," explained to me that it was essential that the training soldiers select a guerilla group that would not impose supplementary or nonaligning goals on the mission. As a lieutenant-colonel told a class of trainees in *Chosen Soldier:* "The indig forces you will work with will have different motivations and values than you. . . . You'll have to find common ground and to pursue common interests with these men; you'll have to win them over" (Couch 2008, 288). Within this logic, an intermediary is meant to help the soldier become an insider without imposing his own political location or aspirations if they are conflicting. Most of the selection process was focused on winnowing out those with extraneous motives. In the exercise for the training captains, for example, there were four guerilla leaders who the soldiers were required to assess. They each had varying motives—from stabilizing the country to vengeance to nonmilitary resistance, and they each had different personalities, levels of trustworthiness, and capabilities. One goal of the training thus was meant to hone the training soldiers' judgment as they assessed the guerilla chiefs.

John explained the challenge of actually working with guerilla leaders in the warzone: "We have to manipulate them [the leaders] to conform to our strategic goals. There are lots of different personalities. We need to get them doing what we want." A trainee who had absorbed this lesson explained: "We have to use emo-intel both to motivate them and some-times to manipulate them." Although the goals of the guerillas were ideally *overlapping* with those of the American military, they were rarely identi-cal. One trainee explained: "Of course those goals will not be one to one. You must make them think that their goals are your goals."

Meanwhile, in this logic, local intermediaries used as human technol-ogy must bear certain bodily costs, in order to protect the bodies of American soldiers. The use of proxies enacts the human instantiation of "virtuous war": "the technical capability and the ethical imperative to threaten and, if necessary, actualize violence from a distance—*with no or minimal casualties*" (Der Derian 2009, xv). Through the use of nationals, the military seeks increasing entry within the warzone without subjecting the American soldier's body to harm. One trainee explained the challenge of convincing their proxies to risk their lives: "The Host Nation dudes might be less willing to fight than you; you have thirty dudes who don't want to go! It's about rapport-building. We need to be brothers to have something to fight for." He continued: "Our biggest responsibility is to train people to do something that helps us. We need to train people to die for their countries, so we don't die for their countries." During one of the training scenarios, one of the guerilla leaders expressed this dilemma, as a disparity in goals might render them unwilling to assume such bodily risk: "We will trust the Americans for a certain period of time before we have to move in another direction." Irrespective of this tension, the goal of the war game was to convince the guerillas to assume such risk, by convincing them that their goals were synonymous with American goals.

Equally essential to this logic, intermediaries must help facilitate the soldier's entry into and access within the local culture and thereby mute his wartime imprint. In the instance of the Pineland trainings, this involved tests that pivoted around both Pineland culture and the cultures of the nonethnic Pinelanders, such as the Iraqis. In her ethnography of the trainings, Anna Simons describes of the training soldiers' relationship to the fabricated culture of Pineland: "Candidates were often made to smoke unfamiliar herbs and grasses and drink vomitus brews. G chiefs routinely made candidates bow down before altars and memorize gibberish chants" (1997, 90). I did not witness these exact practices in my fieldwork; how-ever, Pineland was often described as a culture of its own.

Like in the case with their interface with Middle Eastern culture, the American soldier was coached to mimic his Pineland interlocutors as much as possible. Soldiers were encouraged to participate in local culture and to adopt a culturally relativistic framework when facing what appeared to be innocuous rituals that didn't undermine American operational aims. For example, when the trainees were meeting with one of the more volatile guerilla chiefs, they were required to engage with Pineland mores. The G-Chief welcomed the unit to Pineland: "It's wonderful that the Americans are here. I believe in our customs, and when we have guests, I believe in my drink." At this juncture, he poured them electric red soda in Dixie cups. However, in other simulations, they were offered beer (against military rules to drink) or even a glass of mouthwash. They were required to measure the imperatives of cultural entry against breaking regulations or against their own discomfort.

An American who was role-playing a native Pineland guerilla described the importance of creating cultural difference for the American soldier: "We're told to create weird cultural trends of Pineland, to see if [the trainees] will laugh, if they will or won't partake, you know what will they do if they are given cowdung tea to drink. We would make up stupid rituals, you know awkward, embarrassing, funny things." A white role-player and former military contractor who lived nearby elaborated: "When I'm a G [a guerilla], I make myself like an Iraqi. I try to be *foreign* to the American as a Pinelander. When the Americans show up, we say we hear they're all rich and that they have big cars. We throw out Hollywood stereotypes." Within this conception of cultural difference, although these local intermediaries were mobilized to make cultural entry difficult, they frequently generated the trappings of culture for the trainees. If soldiers could mimic those trappings, they could consider themselves as having gained entry.

Finally, within this military logic, cultural intermediaries harnessed as human technology must both inhabit a stable identity and be capable of stabilizing the identities of others, via reading the human terrain. This imperative exists in uneasy tension in particular with the Green Berets' assumption that everyone in the warzone is potentially a spy until proven otherwise: one major who had been deployed a number of times explained his relationship to local interlocutors of any kind: "You act like you trust someone, but you trust them as far as you can throw them." Employing a local individual in this manner is intricately linked to assuring that the chosen individual can be trusted to read the world and act within the world on the American soldier's behalf. If that person defies the parameters of that use; if the simulation cracks open and the real world pours through and its norms and rules aren't what they expected; if, ultimately, the world soldiers encounter turns out to be unreadable, populated by inexplicable persons and events, then: violence.

.   .   .   .   .

The forty guerillas, all "native Pinelanders," and the twelve Green Beret trainees (who are training the guerilla role-players) are startled to see me stumble into the guerilla camp, belabored by borrowed camping gear in the sluggish humidity.[8] Small offerings begin to come my way. A guerilla helps me set up my tent a tarp for shade; another offers me a bowl of noodles and a fork, which I immediately break. I shift my bowl to my lap, trying to conceal that I am eating only with the fork's prongs; it seems important to impress these guerillas. We sit and chat and the guerillas remain perfectly in character without deviation, even though none of their superiors are around. I have been told in advance that it is important to stay in character, but I am still surprised by the thoroughness of application I see around me.

Before I entered the camp, military personnel assigned me the role of Gypsy. They didn't tell me how they came to the role, or how others in the camp might perceive my presence. Neither the training soldiers nor the role-players know anything about my actual background or why I have come to the military camps; they only know me through the prism of my role, as I present it. In preparation for my role, my military contacts narrated to me:

> "You are from Taylortown outside of Pinehurst. You are the widow of a killed resistance fighter called Joker. Yesterday, a militia called the Sons of Patton came through T-town and started burning houses down, pulling military-aged males out of houses and taking them away, including your son. You should be hysterical. They killed your son in front of you: they cut him up, burned him, and made you eat his ashes. You have nowhere to go, and you want revenge. You go to the base camp to be trained to kill Sons of Patton. If you can work up some tears when you tell your story . . . that would be good."

I remained silent.

One of the soldiers retorted: "Do you take hot sauce on those ashes?"

I flinched.

Another one of them offered: "You better get used to our humor before you go out in those woods."

Once in the camp, I learn that every morning and evening, the guerillas must congregate at the flag to speak the Pineland oath. I learn that I must not approach the firepit until I am invited: in joining a guerilla band, one must slowly earn trust. I eat my noodles next to a guerilla nicknamed "Jackson," the medic for the group. He seems to have a warm and gentle demeanor. He keeps asking me if I need anything, and then he asks me about my background: "What did I do before the war?" The military personnel gave me no guidance on how to reply to this question, having provided me only with the brief character description of Gypsy I just shared. I come to understand that I will be improvising in these woods. I tell him I was studying: I was a PhD student in cultural anthropology interested in understanding war, and then the war fell in my backyard here in Pineland, and now I am interested in Pineland's resistance movement. I decide that it will be easier for me to remember if I make my character as close to myself as possible, and furthermore, this backstory will make it comprehensible to everyone why I am carrying around a notebook and writing notes. As I come to learn, human technology must be stabilized and domesticated as trustworthy (and thus wholly useable) via tests. In my embodiment of Gypsy, the Green Beret trainees erected three major tests to assure themselves of my usability in this regard.

·   ·   ·   ·   ·

The guerillas tell me that I have been selected to go on a six-hour mission, and I must carry an immense amount of water in my backpack. There are two shifts of guerillas leaving in the evening, one at 6 pm and one at 9 pm. I tell them I would prefer to go on the later mission so as not to be out as long. They return and say you're going on the earlier mission: that directive has come from on high. I decide to talk to a man who is role-playing the guerilla chief. Whenever I have role-played in the past, I've had control over what I've done or not done, and I'm feeling the sudden, urgent need to assert this space of autonomy. I met this chief several days prior at a preliminary exercise. I was introduced as a student. I take him aside and apologize to him in advance for exiting my role. I explain that I want to give him a bit of backdrop: I remind him that I am a PhD student from New York, and I'm very excited to be role-playing here, but I am not the most rugged person in the world, and I am the only woman here playing a guerilla, and I am not a contractor, and I want to be able to have a bit of control over what I do—and the ability to say *no* if it seems too physically rigorous for me. His face hardens, and he replies: "Young lady, you've started off on the wrong foot with me. Don't you tell me you're not a contractor like that. If you're here, you'll do everything we do. Or else, you are out of here." I backpedal, apologize, and tell him that I will do my best.

The guerillas help me get ready, pouring icy water on my head in the clinging summer heat and reminding me to spray my boots and clothes with DEET. I follow the men going on the mission where we wait for truck pick-up. We will be driven out to secure a giant field where a helicopter will be dropping supplies at midnight. Because they have not had an opportunity to train me with the AK-47 and the blanks, I do not have a weapon.

I have my first encounter with the Green Beret trainees at this point, who are all on site to help the guerillas with their mission. The American captain, "Sam" (a trainee who is the head of the twelve-man team of trainees) tells me I have to bring a heavy hatchet with me. He rejoins: "You won't need it, but I'm going to make you carry it anyway."

We head out and once we get to the woods, another Green Beret trainee (in charge of this particular mission, but below Sam in rank) realizes very quickly that I cannot keep up with the group and says very kindly to me, "Gypsy, you always stay right behind me." We circumambulate the field to make sure there is no enemy activity, and I push on behind them, drinking a bottle of water with a super-hydration tablet in it. Finally, we get in position and wait in the dark for the supplies to drop. Afterwards, we are driven back to the camp, and everyone who was part of the mission is ordered to the flagpole. I ignore the call, stumbling in exhaustion to tent, am fetched by another guerilla and brought to the flagpole, and finally we are permitted to sleep.

This trial and the subsequent trials I will describe are consistent with the logic of the Green Berets: in wartime, the intermediary must be rendered readable, clear in his or her loyalties and social location, and docile to American aims in order to help the soldier become an insider in the local culture. Yet rather than stabilizing Gypsy in this manner, as the war game progressed, the trainees collectively decided that Gypsy did not make *sense.* In *Dialectic of the Enlightenment,* Adorno and Horkheimer describe how in Enlightenment logics, "anything which does not conform to the standard of calculability and utility must be viewed with suspicion (2002, 3). Reason facilitates processes of locating objects within their appropriate categories, and thereby domesticating them perceptually; previously alarming mysteries, what they call the "insoluble," can henceforth be contained within the letter *x.* Indeed, ultimately anything *outside* the box is perilous "since the mere idea of 'outside' is the real source of fear" (2002, 11). So too, in the war game, the nonsensical must be either tamed or ejected. As a simulation meant to prepare the soldiers for actual war became illogical to the training soldiers, they decided to "break" Gypsy in order to expose her and make sense of her: that is, to know her through her unmaking. But meanwhile, I do not comply, and I risk shaking the entire frame of the game itself.

.   .   .   .   .

"Sparky," the role-player who is the head of my unit of guerillas, asks me to "tell my story." I sit with a group of the guerillas, and in a fairly stilted manner, I begin recounting the almost cartoonishly grotesque story. Sparky starts to hiccup with laughter, which makes me so (genuinely) upset that I retort sharply: "How dare you laugh at a time like this." The other guerillas look at Sparky uneasily: it appears it is inappropriate given the hierarchy for me to challenge him, and the other guerillas defend him. I depart for a tour of the camp but realize that Sparky is in charge of my fate at the guerilla camp, and that I must ingratiate myself with him. I approach him and tell him I would be happy to tell my story again when he would like. He replies that he will fetch the American soldiers, who also need to hear my story.

Via the subsequent initial interrogation, the training soldiers begin to conceptualize me as an outside force, potentially dangerous at worst, and at best, not an uncompliant vessel for American projects. "Sam," the American captain, and his second-in-command lead me to an isolated area. I am becoming genuinely uncomfortable; I tell the story and actually cry while telling it, both inside it and watching myself.[9] In response, Sam says curtly: "When we choose fighters, we only choose those who understand our cause. Certain stories don't make sense. There are people not to be trusted. Two plus two does not equal five, do you understand that, Gypsy?" I answer: "I have just told you a brutal story and this is your reply to me?" He immediately asserts his dominance over me, as he perceives my response to be impertinent. He says: "We are sorry to lose Joker. However, you need to understand that we are in charge and we are not taking in anyone interested in revenge and anyone who is lying to us." I reply: "What is it that you think I am lying about?" He answers that he doesn't think my story makes sense and proceeds to interrogate me about Joker, and then about my studies.

Sam tells me he has studied anthropology too and he wants to know if I have heard of a particular theory. It sounds like a theory lifted out of the *Human Terrain Team Handbook* (Finney 2008), focusing on making the social sciences operationally relevant. When I say I haven't heard of it, he questions my academic credentials, prompting me to ask him if he has heard of Foucault or Bourdieu and would he like me to tell him about their theories? I then tell him that my focus as an academic in "Pineland" is on the Iraqi diasporic community in the region. He retorts sharply: "You're using some big words, Gypsy," then asks if they are Sunni or Shia. When I answer that they are of both sects (on my feet, forgetting that they are Shia in the scenario), he replies: "No, the Iraqis here are Sunni. Do you see how easy it is to catch you in a lie, Gypsy?" However, Sam apparently doesn't know the nuances of the scenario background regarding the Iraqi community in Pineland. I gather myself and answer: "No: they are actually Shia. Shia Iraqis found asylum in the country after 1991 after the uprisings in Iraq." I add (to correct my own prior error) that many of the Iraqis in Pineland come from mixed sect families and the community is more diverse than most people realize. Although he seems vaguely satisfied by my reply, he retorts again: "two plus two does not equal five" and asks me "if I know what happens to traitors in Pineland."

Human technology, to be an effective military tool in wartime, must be maneuvered within Left and Right limits. The American soldier cannot maneuver an individual possessing back agendas, personal or political; any supplementarity compromises the soldier's capacity to minimize his imprint in the warzone. The captain felt convinced that Gypsy had a back agenda, and he became increasingly uneasy when he could not precisely pin it down. During the interrogation, he offered: "We are not taking in anyone interested in revenge," in one stab at understanding the puzzle of Gypsy within the scenario. Yet it appeared that there were other elements that did not make sense to him that he was still trying to decipher. For example, why did Gypsy "use big words"? What kind of foreign irritant was she, exactly? Moreover, there was other excess: Sam saw potent affect within me, teeming over. Indeed, those emotions had been triggered by something very real. I had spent the previous two years interviewing Iraqis who had worked with the U.S. military: they had been pursued by militias and nearly killed before fleeing the country; they were called collaborators by their countrymen. And indeed, despite risking their lives to work for the American military, they were accused of being spies by the American soldiers they worked for. As Sam interrogated me, I felt helpless fury on their behalf.

My position in the role-plays was also getting a bit closer to that of the role-players in some subtle yet eerie levels, as the stressors of the situation compounded. I had been singled out (in the simulation) as a suspicious outsider. But I was not only playing an outsider. I was also nearly the only woman in the group, and I was being interrogated in an isolated location. Moreover, I had very different politics from those questioning me. As the pressure increased, I felt increasingly uncannily enmeshed in the simulations: just as many Iraqi role-players had been. Like them, I was playing a military version of myself (American woman, outsider), but not on my own terms.

·   ·   ·   ·   ·

A group of guerillas and American trainees lead me to three cages at the perimeter of the woods. One holds a goat; one, chickens; and one, a white rabbit. It is swoony hot out, the kind of heat that saps all water out of the body; the pines blur in and out of focus. They have allocated the rabbit to me: "Gypsy, this is yours. Feed it. And then you're going to kill it. We have been kind enough to share our food and water with you out here. You have to do this. We all have to help out so we can eat."[10]

The following day, the soldiers and guerillas lead the animals to the wood-line, where they have dug a pit for unusable portions of the entrails. One of the American soldiers is conducting the "Kill Class" for the guerillas he is training. He explains the proper procedure: secure the goat, slit its neck over the pit, and then proceed with the chickens. "Sparky" calls me forward: "Gypsy, this rabbit is yours. Kill it. Use this stick and hit it in the head." I stand in front of all the soon-to-be Green Berets and all the civilians role-playing guerillas for pay, a sea of men; the pines simmer and contract around me.

Mortified, but unable to stop myself, I begin crying in earnest:

"I can't. I'm sorry."

Someone says: "You have to, Gypsy."

Someone else says: "You can do it."

Sparky tries to give me a fist punch to confirm we are in this guerilla band together. I do not lift my fist.

Someone says: "You have to, Gypsy. How can we trust you? If you can't slaughter an animal for our survival, how can you be a fighter?"

I look up again and the pines swim in the glare. I tell them:

"I'm sorry. I can't. I will clean the animals afterwards if you want. But I cannot do this."

Someone else moves to kill the rabbit. Sparky, his thin body shaking with fury, says: "No, put it back in the cage" and stalks back to the guerilla camp.

I am next escorted back to the animal cages. They have me clean the freshly killed chickens, which I am hoping will serve as a stand-in for killing the rabbit. The flesh beneath their feathers and their organs within is so warm; I sit there in a daze, in the heat, with the knife. I am covered in chicken blood and washing my hands when another American soldier says to me:

> "Gypsy, you need to kill the rabbit now."

> "I'm sorry, sir. Under absolutely no circumstances will I kill this rabbit. Please understand that this will under no circumstances occur."

> "You have to. And you're going to do it. Now. You are with this group and we are being kind enough to feed you and give you water. This is part of your responsibilities here. Unfortunately, you do not have a choice."

> I say: "Listen, I am being perfectly clear with you: it is not happening."

I am again feeling the pines swooning around me, and a bunch of them are yelling out that I have to, and this is the only way for me to be here; and the first soldier says:

> "Okay then, you need to hold it while someone else kills it."

Because I do not protest as fiercely as before, they keep pushing and pushing, and before I know it, the rabbit's feet, tied together, are in my arms. I turn away and begin to cry and then they hit it on the head hard with a stick: I feel it quiver and hear it scream and one of the kinder ones says:

> "It's all over."

But it is not over: the rabbit is still quivering, and they have to hit it again. Afterwards, they bring me over and insist that I skin the rabbit. It is very difficult and very hot out, and I feel dizzy with the ordeal. Thereafter, I am ordered to drain all the blood out of the goat, which is hanging over a pit. Most of the role-players have left, but I am still the focal point as they maneuver me between tasks. Every time I wash my hands, a soldier orders me to do something else: pick up a bundle of organs alight with green flies; scrape the goat meat of fat. Feeling dizzy with the heat, I excuse myself to use the bathroom. The soldier replies that I need to return immediately thereafter for cleanup. He sends me down to carry a large swathe of meat with another role-player, who is instructed to bring me back afterwards.

I begin to understand that my identity and motives are illegible to the soldiers: Gypsy's story and backstory and the facets I have presented about her do not cohere for the soldiers. Some months later, I spoke with one of the guerillas who had interrogated me during the war game, and he explained:

> "In this world, you're a spy until you're not. You did a lot of things wrong. Your story had more than three strikes against you. We had to break you. Everything is fair game in the game. When they see weakness, it is like fresh meat to them."

If someone is unreadable, "outside the box" in the war, that person must be broken. In an infinitely starker example with a brutal outcome, Liisa Malkki describes how Hutus and Tutsis were differentiated, identified, and executed by "body maps," where even colors of tongues were said to be different (1995, 79). She wagers that during the genocide, they would slice open each other's bodies as if looking for a confirmation of identity within the viscera of the body. The threatening entity must be broken into its constituent parts if it cannot be read. At this juncture, the training soldiers might have chosen to expel Gypsy from the war camp—as a suspected spy or simply as a useless entity—allowing the game and the training to churn on. This did not yet occur. Instead, a crack became apparent in the simulation.

.   .   .   .   .

Consistent with their training objectives (a spy cannot be sustained in war), the trainees decide to psychologically break Gypsy, in order to lay out and diffuse the unreadable parts of her identity. Yet, the warscape within Pineland's game takes on its own untamable momentum. As the training soldiers and guerillas continuously try to understand Gypsy's social location in the warzone, I am increasingly interrogated.

"Jackson," the kind medic who had offered me noodles, takes me aside and tells me:

"You see the problem is, Gypsy, we just don't trust you. You come here with this sob story, about being Joker's wife or girlfriend, no one knows really, and none of it matches up. We all already know this isn't the real story. We need you to tell us the truth now. We also know there is an AP reporter named Julie who went missing a few days ago, and rumor has it that she joined up with a guerilla group in order to get that story. We're pretty sure you're Julie. Or else, you're working with [the government police trying to subdue the guerillas]."

I reply: "I am Joker's widow, and my studies got interrupted by the war, and I haven't told you any lies."

He continues: "You also say you're a student researching the Pineland resistance and how we're working with the American Special Forces, but you don't really seem to know what you're talking about. Tell me about the Special Forces, explain the whole selection process."

I begin telling him things I had picked up about the Special Forces, even though the selection process was not relevant to my work. He accuses me of being a farce and not a subject matter expert in my research. At this juncture, Captain Sam approaches us and tells me to give him my notebook. When I push back, he replies:

"I'm not asking you," and proceeds to take it and read the whole thing in front of me, periodically smirking and muttering: "She has like three cover stories," before walking off with it.

After the three tests, I am escorted to a tent for further interrogation. Sam enters and tells me he will be taking my backpack. When I protest, he repeats: "We're not asking." Thereafter, Jackson enters the tent and insists that I "tell the truth now," claiming that they know I am Julie the reporter, or at least not Joker's widow. If they find a picture online of Julie, and it is me, what should they do to me?

I reply: "I am Joker's widow, and I am not lying to you."

Sam reenters the tent and proclaims: "Gypsy, we know you're a liar. I've known it from the moment you came in this camp. Do we need to do this the hard way? I have told you two plus two does not equal five. Your story doesn't make any sense, and we know who you are."

Stunned at this juncture, I ask numbly: "Who am I? I have no idea what you're talking about."

Sam replies: "We saw your ID cards, and we know you're a student at Columbia University and that you studied Oriental Studies at Oxford. It's time for you to start telling the truth."

The interaction is bewildering and uncomfortable: I feel like I've fallen into a vortex that I can't exit.

I reply: "Listen: apologies, but we need to have a frank conversation here. I am just doing the role the major told me to do."

The captain says: "What role? What major?"

The vortex is dizzying; he appears to still be in the scenario, still in Pineland-world, while simultaneously evoking my "real life."

I answer: "I am really confused by this conversation. I'm just doing the role I was given. And yes: I am a PhD student at Columbia. Why does that have anything to do with this?"

And he says: "So you are trying to collect information about us?"

And I say: "Wait, are you still talking about Pineland or the real world?" And he says: "Finally you're being honest with us. We're going to need more of that."

And I say: "Wait, are we still in scenario?"

Jackson says: "Even when we are out of scenario for one minute, we're in scenario right afterwards."

And then I say: "We need to break role right now completely. I have permission to be here."

The tent is charged with our frustration, fury, and bewilderment.

Sam says: "When there's a foreign irritant that is put into the scenario and it makes no sense, it really pisses us off."

Jackson offers: "We can make her the Pineland information officer; she can write pamphlets for us. Are you a good writer?"

Sam answers: "Yes, she is."

Then they decide to just keep me as is, as the Joker's wife, and to tell the students that my story checked out.

He says: "I am going to need to take your phone and every night, you're going to need to read your notes aloud to me. I can't believe they let you in here."

At this point, coincidentally, a sergeant from among the overseeing military personnel (who is outside the simulation, and monitoring it) shows up for a site visit. I approach him and explain what had ensued and ask if I can get my notebook back from the captain (Sam). The captain then speaks with the sergeant to express his discomfort with the situation. I hear the sergeant say: "She has permission at the highest level to be here." Nonetheless, in fifteen minutes, he approaches me, apologizes that it hasn't worked out at this site, and instructs me to pack up my tent.

.   .   .   .   .

In enacting Gypsy, I was reminded of the complexity of my own metaposition in my fieldsite. The introduction of the anthology *Anthropology and Global Counterinsurgency* notes: "The wars in Afghanistan and Iraq have placed new stress on the relationships among anthropology, governance, and war" (Kelly, Jauregui, Mitchell, Walton 2010, 1). The volume insists on the independence of anthropology, critiquing the appropriation of "culture," and its so-called explicator, the anthropologist, by the American military.

The American military's delineation of culture attempts to create a radical displacement of violence onto the Other, while effacing the many forms of American violence that are at play. At stake are geopolitical and neocolonial violence; the psychic violence endured by the cultural Others who are hired; the violence of the imposition of fixed categories upon others for utilitarian ends; and hermeneutical violence (standing as the proprietor of cultural interpretation and rendering other human beings as something more akin to translated texts, or positing that Others can only "be" rather than know).[11]

Meanwhile, as culture is appropriated in this way, anthropology is vacated of its most essential contemporary force from the late twentieth century onward: *critiques of power*, from colonialism to contemporary militarism and war.[12] As I created Gypsy in the woods, I announced within the war game that I was "an anthropology student, writing about war." Moreover, when I later sought to exit the game, I declared that I was actually becoming an anthropologist. The anthropologist is an impossibly charged locus for the American military: alternately seen as a potential interlocutor to help them decipher the elements of the warzone *and* as a potential critic of war, and thus a certain kind of outsider and spy. As the training soldiers and guerillas surrounded me, I in effect became a version of the very complex interstitial individual I had been trying to study. All at once, I was a proxy soldier, an outsider, and an anthropologist—to either immediately expel or to try to court. That is, as a scholar, I was perceived alternately by soldiers as a source of suspicion (a potential spy/critic, carrying around a notebook, and using academic discourses that alienated them) *and* a potential repository of valuable wartime knowledge that might be translated to them, and thus vehicle of potentially helping the soldiers become insiders within Pineland (while in the game), or within the Middle East (as a degreed anthropologist, upon exit). I was also a human being, unusable, and cracking.

On the existential level, human beings are not technologies. They teem instead with the ache of surplus knowledge, grief, dreams, politics, imagination, and refusal, ever controverting the work of the singularly focused tool. The cultural technology is asked to make culture contained and wholly operationally useable, in order to accomplish the work of war. This reification of culture turns it into nothing less than a "corpse" (Daniel 1996, 201).[13] The culture concept totalizes. Meanwhile, as actual objects in the world are continuously subsumed by categorical schema, two plus two are forced to make four. Yet the messy, gorgeous, heartbreaking work of living and dying, of loving and wounding, of trying and getting by—the work of being human across time and space—is not reducible to an equation of any kind.

The U.S. military's cultural training programs, as they still culture into a useable body, as they ask for categorical schema to encompass the seen, are their own invitation to violence. When the real world—its prickly contradictions, its unfathomabilities—come to the fore, training and preparation give way to first confusion, then fury, then violence. When people are militarized as technologies of culture, soldiers are on some level asking for a tidy map of human interiors imagined as other. Do others believe in "freedom"? What will others do under duress? Can others be brought over, to the side of the self? Yet when those maps do not bear out, those human beings must be turned inside out. To break another is to demand they show their insides, to slice open and gut their bewildering bodies and psyches. The military's reliance on cultural intermediaries and proxies instead of American soldiers, the attempt to understand and respect other cultures, was widely lauded as a "softer" way forward in war. This book acts as a counter, calling the violence and omnipresence of American projects to ripple to the surface, as the military hand of the user of the tool is ushered irrevocably out from behind the scenes.

Adorno contended: "The pure unreflective act is violation projected on to the starry sky above. But in the long, contemplative look that fully discloses people and things, the urge towards the object is always deflected, reflected. Contemplation without violence, the source of all the joy of truth presupposes that he who contemplates does not absorb the object into himself" (2005, 90–91). Standing amidst these war markets hawking culture, the fake skies of Pinelandia teeter and fall; the stars burn through.

[Field Poem Fragment]

The wood is long and tall: it's shot with light.
Light makes little lances through the leaves.

Conclusion

THE PINS FALL THROUGH THE PINES

It is an ethical obligation to estrange violence that has become normalized. In some senses, "we now inhabit an army camp" (Lutz 2002); war capitalism is cellular and almost imperceptible: from Lockheed Martin stock in your 401K to mothers posting pictures of their toddler's new rifle and Second Amendment–adorned outfits on a Facebook diapering group to war training camps with fake mosques and fake wounds in them. I have tried in this book to make a permanent state of war and war readiness as devastatingly strange as the haunting rehearsals I witnessed—and their catastrophic dream of American exceptionalism and omnipotence.

One evening, I'm talking with Jack, the Green Beret with whom I had discussed questions of cultural empathy in chapter 1. Between us are one short glass with ice and a lime (the remainder of a gin and tonic) and one empty, tall crystal cup. We've cleared the plates of potatoes and meat, his wife is putting the kids to bed, and the dogs are shuffling under our feet. We've been talking about what happens when things aren't perfectly alike but you assume they are alike or render them so as a short-cut: commensurability. War is full of pragmatic concerns, time constraints, and the need to make use of what you have on the ground. He explains:

"Counterinsurgency turns the crystal into the glass. That's the only route."

"Okay," I say. "So imagine the crystal is an insurgent. But *the glass* is a guy who is being blackmailed, and his family might be killed if he doesn't follow the insurgents. What do you do about the fact of difference? There is a remainder, something about the glass that is different from the crystal."

"You have to turn the crystal into the glass. Or utilize the remainder to your advantage. Get in front of use. Be the first one to use it efficiently or gain value for that use."

"But isn't that so machine-like?" I ask. "Aren't these humans we're talking about?"

"The military believes it is a machine. It functions as a machine. We subjugate your personality; you're a soldier, you need to listen to your elders. It is a machine, and it is surrounded by another machine: the world."

"But how about radical contingency?" I ask Jack. "Can a machine incorporate that?"

"It is extremely difficult," he explains. "It is a problem of resources and time. Can you be Doc Brown in *Back to the Future*? How many parallel universes can you prepare for? If you have one thousand tunnels, based on time, you can only check fifty of them. Unless you can enter a tesseract," he laughs.

Earlier in the evening, he used that word before, and I hadn't known what it meant. I pause the conversation to ask him its meaning, and he reminds me of a childhood book that we both loved: *A Wrinkle in Time*. In Madeleine L'Engle's story, three children leap across time and space to save their father, a scientist researching the tesseract. *A Wrinkle in Time* explains it thus: if space is composed of three dimensions, and time is a fourth dimension, than the tesseract is a fifth dimension—a way to bend the spacetime continuum, thereby connecting two unrelated points on that continuum.

Jack explains,

"In the fifth dimension, you fold time. You can be everywhere at once. You can do anything. Only god-like creatures can enable this."

"How about the soldier?"

"The soldier is mortal, of course. He doesn't have those capabilities. It is ego and ignorance to speak that way. Still, it might be necessitated ignorance. If it helps the soldier to perceive that he's capable of the tesseract, give it to him."

For Jack, it is ego and blindness for a soldier to imagine they have god-like capacities or vision or mobility, that he or she could transcend time or space. The body and mind of the individual alone is impossibly small. But, that person might be multiplied. Jack explains that there are ways for the individual to be multiplied: for example, as a soldier teaching doctrine and training, he ripples outward into hundreds of training soldiers a year. Or, in the case of Unconventional Warfare, which is considered a *force multiplier* by the military, a small group of American soldier bodies are amplified and extended by the local bodies they train and remake. He explains:

> "In this way, an individual can be impotent, or they can be everything if they can multiply and change the universe. He can be the messiah."

This messiah might be seen as a kind of perfect liberal subject: smoothing space and acting on the world—yet without being truly subject to its constraints—and deciphering and maneuvering cultural others who are in the path. Yet the fantasies of war's prefiguration and training meet far more complex friction on the ground. The more a simulation focuses on perfecting itself, the more it becomes sealed off from the actual world. The military cultural training machine is apt to break in the warzone. When the world reveals itself as less tidy and less docile, not only do confusion and sloppiness result, but soldiers may go amok and commit war crimes, engage in arbitrary killings and abuse,[1] or turn the state's war machine to other purposes (like being hired out for illegal extractions or assisting in coups without the military).[2] Indeed, a review conducted in 2020 by the U.S. Special Operations Command, ordered in response to allegations of sexual assault and unlawful killings, noted a problematic culture that focused on combat "to the detriment of leadership, discipline, and accountability" (2020, 7).[3] In the aftermath of a Cultural Turn in part prompted by media outrage to Abu Ghraib torture and prisoner abuse at the beginning of the 2003–2011 Iraq War, military ethical violations and cultural blunders have continued steadily.

Yet, it would be a grave mistake to emphasize here the *breaking* of the military cultural training machine and how its collision with the friction of the world produces violence. Such an analytic would invite the training to incorporate more complexity, to speed up, to yet further meet and mimic the unpredictability of the world. The great catastrophe, and its more frequent outcome, is watching the machine seem to work seamlessly: identifying those who, according to its calculus, deserve to be killed; developing opportunistic friends who can be readily betrayed; and turning human beings into inert bodies of knowledge rather than selves as whole as the American self. The machine going right is the violence all around us that we cannot see.

[Field Poem Fragment]

And why did you work with the Americans?

"I did it to feed my family." "I drove a truck for an Iraqi company that supplied the U.S. military." "I did not know who I was working for exactly." "Who doesn't love a good bootleg Hollywood movie?" Inside our dream of country, a tiny bird sings harshly with oracle.

[Field Poem Fragment]

Out of the sonorous dark came the objects: the bells
beneath the river; the bridge (it is your childhood bridge):

the men and women and children crossing
the bridge on that holy day—

someone yelled: "It is a trap"
and do you know, they jumped that moment

out of their deaths,
the river took them.

From the darkness came the lightdrunk hole
out of the whitehot nerve; came keepers

from the Yusifiyya farm, their bees fanning
the air. Amber-yellow, almost bitter under the sweet,

it can cure your sore throat:
if you eat the eggs of the bees right from the comb

(he explains; have I translated this correctly?)
you will have a very strong heart.

Epilogue

FIELD POETRY

"The wood is long and tall: it's shot with light."

The field poems interlocking the chapters in this book are part of an experiment in using poetry to build anthropological worlds. This coupling was always apparent for me on the level of *subject:* I have written two collections of ethnographic poetry. However, the pairing may also give itself to *form.* This short epilogue is a voyage between theory and poetic craft; it is both a wager on the collaborative possibilities of poetry and anthropology, and a way to further crystallize the difficult world of *Pinelandia.*

"Pins and pines are passing arms. Pins are dropping through the pines."

In the woods after one of the war games, a major announced to me, "the pins and pines are passing arms." This phrase took a little decoding to understand. The "pin," a metonym for the soldier, deflects the military presence into the canopy of trees, seeming to erase military accountability. At different moments and for different people, the pines were a cloak, but they were also described as a wall and a circle. I even wondered if they were a metaphorical lens through which to consider Empire itself, within the context of the war game.

But it was not enough to analyze these metaphors. I'm a poet; I wanted to enact them through craft. Many anthropologists have also written poetry (from Ruth Benedict and Victor Turner to Michel Leiris and Michael Jackson), and bringing anthropology and poetry explicitly in conversation has a rich history. For decades, anthropologists have sought new, hybrid, and experimental forms of representation to bring conceptual frameworks and experiential worlds in conversation. One key genealogy in this regard can be traced to the Writing Culture movement of the 1980s: a reckoning with methods, epistemologies, and the reflexivity of the ethnographer's position, as well as finding new forms of writing to challenge ethnographic realism. James Clifford argued that "the poetic and political are inseparable . . . science is within, not above, historical and linguistic processes" (Clifford and Marcus 1986, 2).

The legacy of this movement is broad, rippling into anthropological thought on representation[1] and wide ranges of experimental anthropological writings.[2] A recent anthology of experimental ethnographic writing, *Crumped Paper Boats*, argued that "the multifaceted character of life leads us to deploy different forms and techniques to engage with different aspects of being," dreaming of "writerly powers and affects that could upturn the political and epistemic status quo" (Pandian and McLean 2017, 15, 23).

One outcome from this genealogy has been a range of new conversations in ethnographic poetry and field poetics, as anthropologist-poets have reflected on the fit between anthropology and poetry, arguing for parallel trajectories in history (Maynard 2002) and potential epistemological cross-pollination (Maynard 2009).[3] Maynard and Melissa Cahnmann-Taylor proposed that "ethnographic poetry" noted the intuitive fit between the modes: "poetry and culture brim with indirection, ambiguity, lacunae, indeed, with downright silence" (2010, 6). Other anthropologist-poets have pointed variably toward such a fit: "the nomadic vagrancy of poetry" (Kusserow in Pandian and McLean 2017), the capacity of poetry to "bring things closer or into focus" (Rosaldo 2013, 105), or poetry as a means of both data and method, a form of research itself as Zani describes in her book: "a field method for recording and analyzing complexity and contradiction" (Zani 2019a, 31). Convergences between anthropology and poetry have also occupied institutional space, for example the Society for Humanistic Anthropology's annual ethnographic poetry contest, ongoing since 1986 (I myself judged between 2014 and 2019); the making of space for ethnographic poetry in the journal *Anthropology and Humanism;* and a more recent contest for anthropological poetry sponsored by *Sapiens* in 2020, as well as an anthology of anthropological poetry in the works coedited by myself and the poetry editors of *Anthropology and Humanism* and *SAPIENS* (2022).

Meanwhile, in the realm of poetics, poets have long wrestled with the border between the poem and the world's current as well as cultural, historical, and political events. As William Carlos Williams writes, "It is difficult to / to get the news from poems / Yet men die miserably every day / for lack / of what is found there" (1955). Still, in the years since, many poets have worked to bring these textures of world and event into the fabric of their poems, in part through documentary poetics.

Philip Metres recently described documentary poetry as "a double-movement both inside the life of the poem and outside the poem." Such a poem refuses to act as a "closed system" instead, engaging with and investigating the world, giving a second life to lost and silenced histories.[4] Within this vein, Muriel Rukeyser's foundational work, *The Book of the Dead* (1938), mobilized her background as a journalist and activist, weaving together witness testimonials, newspaper clips, and court documents, alongside her own interviews, to "extend the document," narrating via poetry the Hawk's Nest Tunnel disaster, and the black migrant workers who died of silicosis. In the years since, other poets have been inspired by this documentary impulse, in particular making use of historical archives. Charles Reznikoff's *Holocaust* (1975) weaves together segments from twenty-six volumes of testimonies of Nazi war criminals. Carolyn Forché, Mark Nowak, Denise Levertov, Jorie Graham, Martha Collins, Natasha Trethewey, Alice Oswald, Philip Metres, Susan Briante, Erika Meitner, Rachel Richardson, Elizabeth Bradfield, Solmaz Sharif, and a range of other contemporary poets might be located within this genealogy of docu-poetry, extending archives within the body of the poem. Some of these poets have worked directly and explicitly from fieldwork and interviews like Rukeyser. In *One Big Self* (2007), for example, CD Wright visits Louisiana state prisons and includes fragments of the voices of incarcerated men and women she overheard. Tarfia Faizullah's *Seam* (2014) incorporates interviews she conducted in Bangladesh of the birangona. In *DMZ Colony* (2020), Don Mee Choi weaves together photographs, transcribed interviews, and imagined monologues, seeking to create what she describes as "geopolitical poetics." Poets have brought in oral history and folklore, often around their own personal histories, such as Rachel Richardson's *Copperhead* (2011), Sarah McCartt-Jackson's *Stonelight* (2018), and Gary Jackson's *Origin Story* (2021). Anthropologist poets have also recently conducted fieldwork for their stand-alone collections of poetry or anthologies, such as Renato Rosaldo's *The Day of Shelly's Death* (2013), Adrie Kusserow's *Refuge* (2013), Melisa Cahnmann-Taylor's *Imperfect Tense* (2016), Ather Zia's *The Frame* (2000), as well as my own two poetry collections: *Stranger's Notebook* (2008) and *Kill Class* (2019).[5]

This work is indebted to all of these genealogies, as it works to crystallize "field poetry." Field poetry first relies on fieldwork—an extended and immersive inquiry into a particular world or set of events, including interviews of the individuals in that orbit. Second, field poetry draws on its specific tools of craft to enact and draw the contours of that world or events. Here, I trace the formal work I did to conjure one temporal, spatial, embodied world from one field site. To this end, my collection of poems *Kill Class* (Tupelo 2019) and *Pinelandia* work as shadow texts of the war games, facing one another.

I contend that with particular resonance, the tools of poetry (structure, syntax, sonics) activate phenomenological experience, offering a more enacted genre—an invitation to enlarge the possibilities of the ethnographic scene. Meeting anthropology's sensitivity to the texture of the encounter, field poetry offers unique tools to illuminate being in a body and being in time: of both living one's own life and imagining and understanding an otherwise. To encounter the field as a poet is to store the sensory and musical alongside the analytic, to perhaps think differently about the rhythms and temporalities of what is lived. In this epilogue, I will perform and analyze poetry's capacities to build anthropological worlds, by looking first at three tools in poetry: structure, syntax and the line, followed by a brief foray into sonics. I will then examine how poetry was a particularly apt tool for *Pinelandia*, showing how field poetics diffuses the reifications of explanation—while reckoning with the essential challenges of poetics in rendering violence without aestheticizing it.

POETRY, FIELD POETRY, AND CRAFTWORK

Structure: Enclosing Pines as Phenomenology and Metaphor

> "The village rises into form between the pines.
> Cows and goats stand in the forest. Flimsy wood
> And storage containers: Muslim quarter   Christian
>   quarter   Assemble/
> disassemble. At a military technology fair in Orlando,
> you can purchase a village in a box.
> Just add people: live inside it for a time."

My GPS didn't work in the woods; I drove in circles until my pulse raced. This phenomenological experience was shared. A major told me about Pineland syndrome: "There would be guys pulling security with a bunch of shrubs. One guy was talking to a tree: *No, Sergeant; Yes, Sergeant.*" Role-players also made comments about these pines: "No way out of here!" was a common joke, gesturing to the wall of trees.

> "The story says we are in the country of Pineland: grassy roads
>   curving in, named
> for longleaf pine, loblolly pine. Sassafras, blackgum, slashpine,
> clethra sharp as pepper, shallowing
> the land's breath."

The pines are something of an exteriorization, a boundary marker of this no-way-out sensation, yet the experience went further for role-players. They also had to stay inside the enclosure of military archetypes. No way out of these trees, *or* out of the constriction of a military logic that categorizes and uses certain bodies and lives, making them substitutable, expendable. Switch the signs in the village, from Arabic to Dari. Switch role-players from Iraqi to Afghan but only if there are enough contractors. Stay in role.

I began to notice this dichotomy of *inside* and *outside* to describe both our physical setting (the pines) and the roles. Military contractors supervising the simulations urged role-players to "Stay *in* the role." One role-player explained to me: "When I'm in role, I stay in the role. You can't talk about anything else." There was an expression used among the role-players, *"aish bi dur": live-in-the-role*. This capacity was alternately praised and viewed with caution. To live in the role was to inhabit it deeply. Yet, a number of role-players had previously been contractors for the U.S. military in Iraq; they'd moved from the war theater to war as theater. To live in the role was thus potentially perilous.

> "Green in here, gleaming
> like being inside a fable
> with stalls of fruit you can't eat"

Once, a former interpreter in Iraq was asked to play an interpreter in a role-play. During the 2003–2011 Iraq War, interpreters were particularly imperiled: accused of collaboration and chased by militias, they had to change their route to work every day or risk being killed. When she was not given a helmet in the simulation (training soldiers were given one), she was distressed. Thereafter, when she was confronted by an insurgent in the simulation, she fainted—a double movement: seemingly falling deeper into role by falling out of it, and into her own past. *"Jeeps grow and grow under the pines."*

I have sought to create a timespace for Pineland that was both entrapped and galloping: to not simply offer ethnographic scene and analysis of a world but to mime sensation within it. My goal was to enter readers into the pines, not allowing them to leave until they are ejected out. However, it became clear that this experience did not mime the sensation of being in the games.[6] We're not always in, but rather we are in and out, in and out, and in. During a break between exercises, I drive with a soldier to a gas station for donuts. Then we drive back in. At night when the games are over, sometimes I drive to the motel with a role-player who used to be an interpreter in Iraq. We pass the so-called real world: McDonalds, Walmart, the Days Inn. We pass a KIA car dealership, and laugh, because KIA is also a military acronym, meaning "Killed in Action." We joke about how these damn woods with their little war keep following us. After a night in the motel, we drive back in.

In order to mimic this timespace in both *Kill Class* and *Pinelandia*, I undertook structural work. I reordered *Kill Class*, dividing it instead into sections. In each section, there is a poem entitled "Driving Out of the Woods to the Motel." Likewise, in *Pinelandia*, both the prose and the in-between poem interludes are thick with pines. In addition, between every chapter, the motel poems loop us back. Every time you think you're exiting the woods, you are circled back in, into this space of Empire where you must live in a role, where critique is hushed. Poetic structure is made into a rhythmic and phenomenological experience: the woods act.

Linebreak as a Unit of Experience

If structure (the macro) is one generative resource, linebreak (the micro) is even more crucial. The tension in a poem is syntactic, created through the release of epiphany as the sentence unfolds over the line. Poet Dana Levin describes the line as "a unit of experience" (2011, 151). It is intrinsically temporal and rhythmic. I'll trace several lines of a poem from my collection *Kill Class* and the first sentence of my ethnographic monograph, *Pinelandia*, to show their interplay. The poem, written first, reads:

> "The fictional country stills
> in the hour's resin. Men glide
> through the pinedark
> into fields of cotton."

In this scene, soldiers must be silent as they covertly parachute into the wargame. The lines release the scene, incrementally. To start, "the fictional country stills" is a unit of experience, where motion ceases. The sentence then breaks, enjambing into the second line reanimating the meaning of each part. The fictional country now stills "in the hour's resin. Men glide." This line makes the country quiet inside time itself. By beginning the next sentence on that same line, "Men glide," the men themselves, via their adjacency, also glide inside that hour. In the third line, they descend, "first through the pinedark" and then "into fields of cotton." The lines slow us, enclose them, change the time of their fall and our witnessing. "It's like everything stopped when we came down," said one soldier.

This poem later became the first sentence of *Pinelandia*. It reads as such: "At this hour, the fictional country is still, and twelve men glide through the dark into a cotton field." In the ethnography, as in the poem, it is not enough to just say the thing. There are other tools, rather, to perform it. I wanted my readers to feel the eerie, almost-sweet vertigo in the soldier's descent, the late hour a sort of dark honey around us, as they tumble into the cotton field. There, such softness, the cotton bolls under their boots. The prose sentence borrows the affect, rhythm, and imagistic arc of the poem: we begin our entry with time ("this hour"), then move into the strangeness of place ("the fictional country"), and its state ("still"), and each of these clauses slows the reader, before we arrive at the key action in the sentence: "twelve men glide through the dark." Their bodies fall through the dark like the *l*'s in the words, and the sentence itself brings us through their descent all the way down to the field where they land. That moment of secrecy and almost terrible softness: this calm before the war begins. In this way, an ethnography contains the ghosts, the trace-structures of the poem.

Sonics and the Ear

While anthropology's Writing Culture movement did little to theorize the use of poetry itself in the discipline, the introduction to that book perhaps anticipated it, noting that anthropology's key tool was *listening*. Still, perhaps the ethnographer has thought less on how to translate the listening ear itself *via writing*.[7] In poetry, sound and rhythm map out time, experience, and revelation by imprinting patterns and departures in the sentence upon the listener. What Robert Frost calls the "imagining ear" in 1915 opens towards "a plot line of sounds, of feeling" (Voigt 1999, 108) through "the human body as its medium" (Pinsky 2014, 9).[8]

> "The **wood** / is **long**/ and **tall:**/ it's **shot**/ with **light.**
> **Light** makes/ **little**/ **lan**ces/ through the **leaves.**"

These lines I wrote one day when pinpricks of light came through the woods. I wrote it on a day that felt lighter: role-players were joking, talking about how the war game was only a game. Leaving the woods seemed easy, the light glinted through. The first line is in iambic pentameter (five units of unstressed-stressed syllables): a cadence like walking, and this case, seeing one's surroundings anew. The second line begins three trochees (stressed-unstressed) followed by an anapest (two unstressed syllables followed by a stressed one): the light pushes through on the stressed syllables.

The craftwork here relies on enactment—or mimesis. *In poetry*, mimesis is when form mirrors or enacts content. Mimesis as a concept has also crisscrossed philosophy and anthropology, and it has been variably understood: an imitation; a replica; a representation; a way for subject to dissolve into object: wave your arms, become a windmill. Michael Taussig uses the concept differently over time, first noting its subversive colonial capacities (1993) and a decade later conjoining it with poetics. Of Sylvia Plath's book *Ariel*, he describes how mimesis occurs when "her poems enter into the things they refer to, and take me along with them" (2001, 305). That is, the reader seems to *die-into* the poem: a collapse between self and other, subject and object, a sorcery with words. Describing Plath's poems, Taussig conjures "a mimetic burnout of metaphor fused at white heat into the referent" (2001, 315).

Yet this proposition of dying-into anything—a place, a time, a lifeworld, or even a representation—is risky. We would not want to *go-native* inside the poem: fantasizing of being-engulfed by the object, while paradoxically engulfing it. This is a very different act from trying to understand and marking our separateness. Poetic mimesis (when form mirrors content) may generate a spell. The ethnographer who uses poetics might most ethically formally mime the object while then reinstating the subject, making both perceiving self and the act of constructing a text, legible. These are, of course, old lessons in anthropology. I resummon them amidst the seductions not only of mimesis, but of poetry's tools that have rarely been considered in anthropology.

In my work, I work to note that we're in a *made* object, a poem that knows very well it is a poem. Ethically, politically, epistemologically, we cannot, we must not, die-into the woods and into these painful wartime experiences of others. For example, one poem draws a mimesis that marks our distance from the thing-itself: we can only go *in* up to a point. We are, after all, inside a poem, which must be marked: "Whichever way you turn, soldiers / dream-walk in and out of the poem. The dark / is sweet: a nerve inside a tooth."

Amidst this focus of poetry, we must be wary of the seductions of beauty, of language's magic, particularly when writing about violence. As

Fatima Mojaddedi states in the introduction to Anand Pandian's *A Possible Anthology*, of her own fieldwork in Afghanistan: "What does it mean to privilege metaphor, or the magic of words in ethnography, in a place where people can die for speaking metaphorically?" (Pandian 2019, 12). She's right. And yet, and still, metaphors and the magical work of language so often structure the imagination even when they are unspeakable to power. Metaphor, when used to think with, help us understand specific worlds, and their nightmares and forms of entrapment. The ethical obligation of the use of aesthetics here is to remember always that violence is ugly. But poetics might be a crystalline point to show its ugly and horrifying granularities (rather than through the obfuscations of academic writing), and our obligation here is to particularity. I will next argue that the poem might indeed wield a critical capacity *against* at least a certain kind of representational violence, looking at my own case study in the training camps. For representational violence, representational redress.

FIELD POETRY AND THE MILITARY INDUSTRIAL COMPLEX

In chapter 4 of this book, I described how countries, cultures, and people are turned into interchangeable, replicable cultured-bodies and archetypes—reifications circulating through a war market. This kind of reification takes any phenomenon as a given, thereby turning complex, shifting processes into knowable, containable, and homogenous objects. Indeed, Adorno proposes that any phenomenon that is *not* reified "cannot be counted and measured, [and] ceases to exist" (2005, 47). Wartime requires counting and measuring: knowing an object in order to control it. "This is the Middle East," the trainings teach. "This is what Arabs do when they're relaxing. This is what they do under pressure." "This is culture." While the role-players largely sought to play their roles seamlessly to retain their lucrative contracts, they did not view these reifications as accurate portrayals. Their reactions to these reifications brought an uneasy energy to the mock villages.

In chapter 5, I described the Crying Room, a scenario where U.S. soldiers had inadvertently crushed a child with their tank and now must pay their condolences, while ideally acquiring intelligence about the family's connections to the militia. The role-players are asked to repetitively embody military reifications of Iraqi women: women who grieve with theatrical flourish but remain docile to the U.S. presence in their country; women willing to betray their brothers. As they mechanically grieve this reduction of wartime Iraq, they weep, screaming, hitting their thighs; and very occasionally, they make jokes in Arabic and laugh. I argued that this laughter acts as an interruption to the reification: a very momentary fracturing of the military's "Iraq."

What, then, is the role for poetry here? It has always been my intuition that poetry as a form intrinsically resists reification. As form, the poem is not an immediately consumable, knowable, or homogenous object. Rather, a poem resists such containment: through its multiplicity and potential variable interpretations; through its strangeness, its silences, and the fact that it is largely not market-driven. A masterful poem remains resistant— that is, strange and splendid even after reading it many times. I tell my beginning poetry students: don't try to find out what the poem *means*. Rather, let's discover what it *does* inside the body of the recipient. For example, the play of low vowels in a poem might ache open an anticipation of loss. Or if a regular meter is disrupted and the poem cracks, this might create a sensation of doubt. In *How Poems Get Made*, poet Jim Longenbach describes a good poem as "infinitely repeatable, richer over time" (2018, 12), noting elsewhere that a poem conveys knowledge "only inasmuch as they refuse to be vehicles for the efficient transmission of knowledge" (2004, inside flap). In his essay "The Storyteller," Walter Benjamin describes "the story" in a similar way: a compressed, mysterious kernel, which must not be "shot through with explanation," which must not be immediately consumable ([1968] 2007, 33). Likewise, the poem asks us to engage with it again and again. Like the vividness and complexity of a country or a person.

As a form, the poem is perhaps in some ways *the opposite* of the signifiers that litter war training camps, which purport towards a single way of being and knowing. In the United States' military industrial complex, Arabs are impatient. Sunnis hate Shiites. Men oppress women. There are bad guys and good guys. There is an equation where everything is knowable if you get the right piece of intelligence. All contingency can be incorporated into a storyline.

Conversely, the inexhaustibility of poetry is in part because the poem is not exactly the *what* (the subject, the meaning, the *X*) but it is rather the *how* (music, the experience of the song). As Jim Longenbach explains: "No one reads Keats ode 'To Autumn' to be reminded that in September leaves turn colors and fall from the trees; even if we know the poem by heart, we savor the experience of the poem's language as it unfolds in time, luring us forward (2018, 11–12). We tumble through its ripe sounds as they diminish: "Season of mists and mellow fruitfulness, / Close bosom-friend of the maturing sun." This Longenbach describes as *lyric knowledge*, the pleasurable rediscovery of what we might already know. The reader is asked to be an experiential co-creator of the poem every time they may come to it. As Maynard and Cahnmann-Taylor have pointed out, both anthropology and poetry veer towards the unsayable, anthropology describing "what goes without saying," and poetry's strength at articulating the inexpressible (2010).

Against the reifications of the mock village, I wrote a long poem inspired by an inventory of titles of books: the actual books themselves had been lost in the 1258 Mongol sacking of Baghdad. I wrote the poem for an Iraqi role-player friend who was also a writer and who loved al-Mutanabbi Street, the beloved street of book markets that was bombed in 2007. The titles of the books were magnificent, mysterious and painful in the absence of the books: there was *The Drawing of Lots* by Ibn al-Mutahil, *Coming on Objects Unexpectedly (verse) by unknown*, and many others. I broke the poem into sections, each prefaced by a question including a title of a lost book. The poem contained imagery from stories and dreams I'd heard during field-work; experiences I'd had with my interlocutors; interview quotes and occasional lines from the Iraqi poet, Badr Shakir Al-Sayyab, a poet many of my interlocutors could recite. These were braided with meta-questions about my own process of translation (as a poet and an anthropologist). What follows is the final section of that poem, entitled: "And who are you now, on the other side, *Coming on Objects Unexpectedly (verse) by unknown?*":

Out of the sonorous dark, came the objects: the bells
beneath the river; the bridge (it is your childhood bridge):

the men and women and children crossing
the bridge on that holy day—

someone yelled: "it is a trap"
and do you know, they jumped that moment

out of their deaths,
the river took them.

From the darkness, came the lightdrunk hole
out of the whitehot nerve; came keepers

from the Yusifiyya farm, their bees fanning
the air. Amber-yellow, almost bitter under the sweet,

it can cure your sore throat:
if you eat the eggs of the bees right from the comb

(he explains; have I translated this correctly?)
you will have a very strong heart.

The poem is composite. In it are allusions to Sayyab's poem, "Death and the River," with bells beneath a river. That river melds into a story my interlocutors told me about the 2005 stampede of pilgrims while on a bridge en route to the Shia shrine of Imam Musa al-Kazim. There is also the story of an Iraqi beekeeper who desperately wanted to one day return to his farm. It ends with my own query: *Have I translated this correctly,* that compulsion that drives both writing an ethnography, and in some sense also writing an ethnographic poem—even one that is much more dreamily and hybridly tethered to its referent.

I don't know what the poem *means.* I know that when I wrote it, I felt haunted by a conversation I'd had with a number of Iraqi role-players who had become friends. After the war games (spaces that had very little to do with Iraq for them), we would go to lunch at someone's house and watch the Iraqi news as the Islamic State rose to power. The news would be on all afternoon, while we cooked and ate, and someone would eventually sigh and say *Iraq is over.* Thereafter, someone else would add, *Maybe Iraq isn't over. Give it twenty years. Maybe there will be an Iraq again.* Then we would sit around and they would tell me stories about Iraq: beautiful and terrible things they remembered. I wrote the poem as a dream-reply to those conversations: "out of the sonorous dark, came the object. From the darkness, came the lightdrunk hole / out of the whitehot nerve."

A poem (perhaps more compressed and mysterious) lives alongside an ethnographic monograph where there is room for more *explaining.* Who was at the bridge? Who was the beekeeper? And on. Still, on its own, there was something in the poem of what I learned about what my interlocutors had lived in Iraq: the beauty and fear, the honey and sting, the goneness of what it was and the imagination of an otherwise. The poem is just one representation of Iraq—one glimpse from my own ten years of engagement with the country and its diaspora. Still, I wrote it to hold it up as a counter to the military's glossy plastic Country Card about Iraq, the erasures of its bullet points. If American practices of war-making and cultural training act as a death that reify person and place, poetics might be one kind of jolt back alive.

As anthropologists, we so often pivot between ethnographic scenes, vignettes, and anecdotes and various forms of explanation and analysis. Poetics offers a different approach to a scene's evocation of the ethnographic present: the text *does the scene rather than says it*. Additionally, as field poetry works to crystallize a moment of experience, it might embed an epiphany in language itself, rather than extending a separate explanation, which risks distancing from the experience, sensation or affect described.

Shaking off the plaster cast first of military representations, and then of anthropological explanations, we are left with a spiral into experience, a more acute kind of knowing and becoming through that knowledge. In contrasting the starkness of the instrumental line, Adorno instead offers the spiral, his idea of tenderness itself: rather than extracting something from the Other, we are approaching their life—its vivid, complex, changeability—with the care we would dream of in approaching our own.

[Field Poem Fragment]

You asked the part
of me I kept hidden. It was every
softness I didn't give them,

the life awake,
whole,
trembling.

Notes

1. Several Green Berets I interviewed described T. E. Lawrence's *Seven Pillars of Wisdom* as "doctrine." The General Purpose Forces also began to increasingly read Lawrence amidst the 2003–2011 Iraq War counterinsurgency trainings. A Green Beret captain trainee used this line with his soldiers during one of the Unconventional Warfare trainings that I witnessed.

2. This kind of training begins for captains in the early stages, and for all Green Berets in their culminating exercise, Robin Sage. Pineland is the site for rehearsals by the Special Operations Forces Community (the Green Berets or Special Forces, as well as the other two branches: the Civil Affairs and Psychological Operations). They rehearse a range of stages of war, from Unconventional Warfare to counterinsurgency trainings in which soldiers interact with cultural role-players to psychological operations to later stages of conflict and civil affairs. See other studies of the Green Berets' training and exercises: Anna J. Simons's *The Company They Keep: Life Inside the U.S. Army Special Forces* (1997); Dick Couch's *Chosen Soldier: The Making of a Special Forces Warrior* (2008); Tony Schwalm's *Guerilla Factory: The Making of Special Forces Officers* (2013); and Robin Moore's *The Green Berets: The Amazing Story of the U.S. Army's Special Forces Unit* (2016).

3. Pineland has been expanding. In the early 2000s, it was reported at 4,500 square miles and by 2009, at 8,500. *Special Warfare* 24, no. 2 (March-April-May 2011).

4. Since 1952, the Green Berets have trained for Unconventional Warfare in different settings across the United States (West Virginia, Georgia, North and South Carolina) with differently named exercises (previously, "Gobbler Woods" and "Cherokee Trail"). The creation of Pineland in North Carolina dates back to 1974, and the Green Berets' culminating training exercise was renamed Robin Sage after it moved near Robins, North Carolina. Training operations span the North Carolina counties of Alamance, Anson, Cabarrus, Chatham, Davidson, Guilford, Hoke, Montgomery, Moore, Randolph, Richmond, Rowan, Scotland, Stanly, and Union at the time of writing.

5. Trainings were eight times a year at the time of research, however changed to six times after 2014 and may shift again. U.S. Army, "Robin Sage Exercise to Run in North Carolina Counties March 9–20," March 4, 2013, https://www.army .mil/article/97667/robin_sage_exercise_to_run_in_north_carolina_counties_ march_9_20; Proceedings of the Annual General Donald R. Keith Memorial Conference, West Point, NY, May 2, 2019, http://www.ieworldconference.org /content/WP2019/Papers/GDRKMCC-19_49.pdf.

6. Contractors (often either veteran owned or staffed) compete for Department of Defense and Special Operations jobs to run portions or the sum total of any given training program, including the creation of the facilities and décor (realistic contemporary operational environments) and managing the hired employees (the role-players and any other subject-matter experts or linguists). In addition to the presence of hired contractors, locals in the area regularly grant the DOD use of their land or facilities for training scenarios. See Woytowich (2016) for more details.

7. Quotation from "The Relevancy of Robin Sage," www.soc.mil/SWCS /SWmag/archive/SW2902/Robin%20Sage.pdf.

8. In *Bomb-Children: Life in the Former Battlefield of Laos* (2019a), Zani makes use of fieldpoems as both data and method. I agree with Zani's depiction of fieldpoems as a process of "poetic inquiry" (following Prendergast 2009, xxii), making room for ambiguity and uncertainty. However, I do not see poems as data but rather as a phenomenological means of activating experience in the reader.

9. See Zaynab Saleh (2021, 14) for an excellent chronology of U.S. imperial interests in Iraq since 1958, after the fall of the monarchy, ranging from attempts to safeguard American access to Iraqi oil to working to rearrange geopolitics in the country (i.e., deter Iraq's embrace of communism in the Cold War or support regimes to create American versions of "regional stability"). It is essential to note the long history of imperial intervention in Iraq, and to not create an approach of "post 2003 rupture" (Ali 2018).

10. See also Derek Gregory for an incisive description of how the GWOT is a key mode of what he calls "the colonial present": "The war on terror is an attempt to establish a new global narrative in which the power to narrate is vested in a

particular constellation of power and knowledge within the United States of America (2004, 15–16).

11. Using virtual reality such as satellite imagery and street data, the U.S. Army has worked to develop Synthetic Training Environments, having built virtual versions of cities such as San Francisco, New York, and Las Vegas. See www .weforum.org/agenda/2018/04/soldiers-are-training-in-virtual-environments-generated-from-real-cities; and https://usacac.army.mil/sites/default/files /documents/cact/STE_White_Paper.pdf for more.

12. This was particularly the case under President Trump, when militarism was used to suppress demonstrations and unrest in the United States.

13. "The Biden Doctrine: The U.S. Hunts for a New Place in the World," *Financial Times*, Sept. 3, 2021, www.ft.com/content/2a88ac0b-d3d7-4159-b7f5-41f602737288.

14. A range of programs facilitated the entry of Iraqis into the United States. In 2007, at the height of sectarian violence, only six thousand Iraqi refugees had been admitted to the United States. Larger-scale processing was initiated at that juncture, focusing on the especially vulnerable, such as those impacted by sectarian violence or affiliated with the U.S. government. Between 2007 and 2013, 84,902 Iraqi refugees were admitted via the U.S. Refugee Admission Program. The Special Immigrant Visa Program, which stipulated the admission of five thousand affiliated Iraqis each year between 2008 and 2012, rarely met quotas, in fact issuing fewer than five thousand visas during the entire period. More recently, see the widespread media reports on the case of Afghans left behind, e.g., www.independent.co.uk/asia/south-asia/afghanistan-interpreters-taliban-home-office-b1910410.html.

15. See my article, "Imperial Mimesis: Enacting and Policing Empathy in U.S. Military Training," *American Ethnologist* 45 (4): 533–45, for an extensive discussion on the implications of weaponizing affect, in particular "empathy," as well as portions of chap. 1.

16. Anthropology's imbrication in colonialism has been well charted (see, especially, watershed texts: Asad 1973; Stocking 1991; Pels and Salemink 1999, and many others since the discipline was reconfigured through a postcolonial turn). More recently, anthropology has sought to "decolonize" the field (Harrison 1991; see also the varied Decolonizing Anthropology series via Savage Minds, edited by Carole McGranahan and Utma Z. Rizvi: https://savageminds .org/2016/04/19/decolonizing-anthropology.

Otherwise, many scholars have focused on how expert and cultural knowledges are used to manage populations in wartime and how anthropology has been implicated (Gregory 2008; Network of Concerned Anthropologists 2009; Kelly et al. 2010; González 2009; Price 2004, 2008, 2016), critiquing the use of culture as a "weapons-system" (Davis 2010) and turning to unsettling precedents such as Ruth Benedict's 1946 wartime ethnography of Japan, *The*

Chrysanthemum and the Sword: Patterns of Japanese Culture, commissioned by the U.S. Office of War Information to help the American military predict the adversary's moves. See also Price in particular, for a genealogy of entanglements and interactions between American anthropologists and U.S. military and intelligence agencies, from *Anthropological Intelligence* (2008) focusing on World War II, to *Threatening Anthropology* (2004), focusing on FBI surveillance of anthropologists during the McCarthy era, to *Cold War Anthropology: The CIA, the Pentagon, and the Growth of Dual Use Anthropology* (2016) and its analysis of how anthropology as a discipline has been influenced by U.S. imperialism, as well as the role of CIA funding. Regarding the crucial entanglements between Area Studies and militarism, and Area Studies' intrinsically political history as a "child of empire," see Said (1978) and the anthology *Learning Places: The Afterlives of Area Studies* (2002). Rey Chow (2010, 14) describes how in tandem with the Cold War, regions of interest in the U.S. became "areas to be studied, these regions took on the significance of target fields—fields of information retrieval and dissemination that were necessary for the perpetuation of the United States' political and ideological hegemony."

17. In 2007, the American Anthropological Association's Executive Board condemned in particular the Human Terrain System (where social scientists were embedded with military units) as the "unacceptable application of anthropological expertise" (AAA 2007), noting the potential harm to human subjects and the impossibility of informed consent. Decrying the reification and weaponization of cultural knowledge, many anthropologists supported the Board's statement.

18. Thank you in particular to Joe Masco for conversations about permanent American war and the urgency to estrange that mode. See also Joshua Reno's excellent discussion of "permanent war readiness" (2020).

19. American military and civil defense live action simulations in mock villages are hardly new, among either Special Operations Forces or General Purpose Forces. However, the post 9-11 widespread contracting of "cultural role-players" within such spaces represents a particular historical moment in the U.S. military-industrial complex. Although Vietnamese role-players were occasionally hired in Vietnam War trainings, U.S. soldiers themselves more frequently played guerillas, and most U.S. trainees were not exposed to the program (Westheider 2007, 61). Later Combat Training Centers most typically employed role-players drawn from military manpower, before the use of cultural role-players.

20. The CTCs developed in tandem with new counterinsurgency training exercises; namely, virtual exercises in programmed training environments, where trainees both face kinetic threats (e.g., improvised explosive devices) and have cultural interactions with Middle Easterners. In recent years, warfare, drones, and digital war games have increasingly become entangled, potentially narrowing the gap between the real and the virtual (Der Derian 2009) and cre-

ating commuter warriors (Gusterson 2016). Meanwhile, the army began recruiting through America's Army, a virtual education center featuring interactive games that blur the line between war and entertainment; it is the first large-scale gaming technology used for U.S. Army recruiting. Also used for this purpose are a range of other games available on the online multiplayer gaming service Xbox. Civilians who play war games in virtual environments may be good candidates for the Army, forming a continuum of casual players to "virtual soldiers" (Allen 2017, 148).

21. Meanwhile, the Special Forces have been training using simulations since the 1950s, drawing on both military manpower and local individuals living in the environs of the trainings as role-players, and later, Montagnards, who fought alongside the Green Berets during the Vietnam War and then emigrated to the United States. The phenomenon remained relatively low-budget and contained in the Special Forces community prior to 9-11.

22. A phased plan to increase Special Forces graduates was introduced in 2002, "to grow from a ten-year average of 350 active duty graduates enlisted per year to 750 for FY 2006. However, in 2005, the SFQC produced 790 new Green Berets—exceeding the goal a year early," Department of Defense Authorization for Appropriations for Fiscal Year 2007, Hearings Before the Committee of Armed Services, United States Senate, 109th Congress, Second Session on S. 2276, p. 1163, www.google.com/books/edition/Department_of_Defense_ Authorization_for/6Z4sLqyStv8C?hl=en&gbpv=1.

23. I engaged in diverse methodologies, relying on extensive "participant-observation" and conversations as well as over ninety interviews with a combination of U.S. military personnel and Iraqis who had worked with American soldiers or companies. Interviews and conversations with Iraqis were largely conducted in Arabic. Occasionally conversations occurred in English with former translators who were fluent in English.

24. Combat missions in Iraq actually continued in different iterations for many years longer, i.e., with U.S. soldiers assisting Iraqi forces. In 2021, President Biden stated that combat missions would conclude by the year's end, but he did not note if he planned to reduce the 2,500 troops on the ground. See https:// apnews.com/article/joe-biden-government-and-politics-middle-east-iraq-islamic-state-group-9397d9996703d7416f857165072a0a05.

25. I describe the post 9-11 Cultural Turn, the military's focus on cultural literacy in wartime trainings, in the introduction and in chap. 1. The term's genesis came several years into the 2003—2011 Iraq War. In 2007, Patrick Porter described a "Cultural Turn" in the military's field of counterinsurgency studies in "Good Anthropology, Bad History: The Cultural Turn in Studying War" (2007). The term entered the lexicon among anthropologists after Derek Gregory's essential article, "The Rush to the Intimate: Counterinsurgency and the Cultural Turn in Late Modern Warfare" (2008).

26. "Phase Zero" entered the war lexicon in Charles Wald's 2006 article in *Joint Force Quarterly*. The strategy was elaborated later that year in the *Quadrennial Defense Review*. "The QDR argues that victory in the 'long war' against terrorism requires bolstering weak and failing states so they can better defend their borders and territories and eliminate 'ungoverned spaces' hospitable to America's enemies. Accordingly, the U.S. military should expand training of foreign security forces and cooperate with U.S. civilian agencies in engaging developing countries." See: www.cgdev.org/blog/phase-zero-pentagons-latest-big-idea.

"Left of Bang," a new permutation of this same concept (Phase Zero) entered the war lexicon officially in 2012; however, it was used earlier colloquially. According to an article by military intelligence analysts Michael Flynn, James Sisco, and David Ellis, referred to as a bible by some in the national security community in this period: "Left of bang, policy options are more numerous, costs of engagement is lower, and information flows more freely to more actors. After bang, options decrease markedly, the policy costs rise rapidly, and information becomes scarce and expensive" (Flynn, Sisco, and Ellis 2012, 13–14). Patrick Van Horne and Jason A. Riley also describe how the Marines' Combat Hunter Course was designed in 2007 with a "Left of Bang" framework (2014, 18).

27. The Special Operations Forces *Unconventional Warfare Field Manual* (2008) defines guerilla warfare: "Military and paramilitary operations conducted in enemy-held, hostile, [or denied] territory by irregular, predominantly indigenous forces."

28. See Kipp, Grau, Prinslow and Smith (2006).

29. For Jeffrey, the strategy was intrinsically flawed as it could only execute the first two (the explicitly military aspect) of counterinsurgency's supposed three legs ("clear, hold, build"): it was less able to execute "reform and reconciliation," the final building component (2015).

30. Further, while President Obama worked to "tighten the metrics" on drone operations, he also ultimately heavily relied on drones in nonbattlefield settings. In addition, Obama used a contested method of counting civilian casualties, likely misrepresenting deaths in statistics. See https://thehill.com/policy/defense/570280-biden-likely-to-lean-on-drone-warfare-in-afghanistan.

31. For those who enact remote war, this is certainly more complex, and personal accounts of some drone operators have described significant amounts of stress and trauma.

32. Kali Rubaii (forthcoming, *American Anthropology Special Issue: War on Terror*) describes the "Iraqicizing of violence" in the later stages of the War on Terror, the strategy to "maneuver responsibility for violence onto Iraqis themselves"—through indirect rule and the training of indigenous partners. She notes that Iraqicization only offered the "pretense of Iraqi sovereignty, which was important to the Coalition (CPA), who risked accusation of war crimes." Conversely, rather than fostering actual Iraqi sovereignty, this strategy entrenched

sectarian rifts (as the United States appointed Iraqi politicians from Iraqi opposition groups in exile). Further, Rubaii notes that this strategy "also fueled a discourse about the moral authority of U.S. intervention as a civilizing global force."

33. For more on Trump's escalating involvements, see: www.thebureau investigates.com/stories/2018-01-19/strikes-in-somalia-and-yemen-triple-in-trumps-first-year-in-office; and www.newamerica.org/international-security/reports/americas-counterterrorism-wars/the-war-in-somalia; and https://chicago.suntimes.com/news/2019/5/8/18619206/under-donald-trump-drone-strikes-far-exceed-obama-s-numbers. For Biden, see: www.nytimes.com/2021/03/03/us/politics/biden-drones.html; for discussion of the emerging Biden doctrine, see https://newrepublic.com/article/163744/biden-doctrine-united-nations-speech.

34. Laleh Khalili rightly notes that "culture" and "civilization" are euphemisms for race, and further points to the paradox of a mode of warfare deemed less violent (and veiled via liberal and humanizing language), further legitimizing war itself as a necessary political intervention.

35. Catherine Lutz urged anthropologists to do what they do best, to mobilize a "polycentric approach to knowledge" to see around the "imperial, mission-centric approaches" of other disciplines like political science and economics, and to harness ethnographic methods to listen closely to "both those who benefit and those who suffer in the imperial relationship" (2006, 598). This call has been taken up in recent years as ethnographers excavate the daily contours of Empire (McGranahan 2010; Dennison 2012; Simpson 2014; Vine 2015; McGranahan and Collins, 2018).

36. Stoler describes the "disassemblage" as a way of severing interconnected events and relations, such as in the "U.S. imperial script: that Samoa is not related to the Philippines is not related to Nicaragua is not related to Iraq" (2018, 478).

37. My notion of Empire is inspired by several other sources. I am partly influenced by Hardt and Negri's notion of *Empire* (2000, xii–xv) as a "global form of sovereignty," in which the "object of rule is social life in its entirety," not simply a mode for "manag[ing] a territory and a population but creat[ing] the very world it inhabits." Within this "deterritorializing apparatus of rule," Hardt and Negri contend that the United States occupies a "privileged position." Further, they propose an essential paradox with Empire: wherein its "practices" are bloody but its concepts are dedicated to peace. I am less analytically interested in Hardt and Negri's insistence on a qualitative break with "imperialism," a before and an after. In the preface to the anthology *New Imperialisms*, vol. 1, Maximilian Forte notes that "new imperialism" might be understood as "liberal" or "humanitarian" imperialism, or "military humanism," but "it is not all that 'new'" (2010, xi). Meanwhile, Ann Stoler (2016) proposes that instead of viewing discontinuities, we examine subtle and tenacious "imperial durabilities," where various zones of global precarity "are intimately tied to imperial effects" (3).

38. Rey Chow brings war, racism, and knowledge production together as a crucial trio as well in her examination of Area Studies and producing the world as a target (2010, 13). Chow notes: "the pursuit of war—with its use of violence—and the pursuit of peace—with its cultivation of knowledge—are the obverse and reverse of the same coin, the coin that I have been calling 'the age of the world target.'" In my case study, this blurring extends even further, as knowledge production is inextricable from active warmaking itself.

39. My conception of militarization is likewise inspired by several sources. The military historian Richard H. Kohn defines militarization as a set of processes that show "the degree to which a society's institutions, policies, behaviors, thought, and values are devoted to military power and shaped by war" (2009, 178). Regarding militarized spaces, Catherine Lutz delineates a paradigm where war is understood as the "health" of a country (2001). She describes how the United States has remained in a state of permanent readiness for war since World War II, noting that the country's culture, economy, politics, forms of entertainment, moral compass, and so on "have all been shaped by military institutions that developed as instruments of conquest and control of Native Americans in the so-called Indian Wars, which began in the late 1700s and continued for more than a century." In discussing the militarization of culture, Roberto González describes how even human connections are infiltrated by the military-industrial complex. Most to the point is the militarized human being, which is perhaps an extension of these other logics. The person is militarized in that they are turned explicitly into a tool for the military, and that their actions are in some way co-opted by military institutions.

40. See chap. 1 for further elaboration on the entwinements between sentiment and power, and the bringing together of sentimental means and operational ends.

41. Recent scholars have focused on how soldiers have become technologies of war-fighting and rehabilitation (in part through examining how they become composite—with armor, prosthetics, and the like), see MacLeish (2015), Jauregui (2015), and Wool (2015).

42. Cultural practice is transformed and repackaged both into a source of economic value (as is described by Julia Elyachar in *Markets of Dispossession*), and into a source of *military value*. Elyachar (2005) shows how, amidst the explosion of Neoliberalism in Egypt, cultural practices that had been deemed "backwards" were turned into new forms not only of social capital but also of profit (which led, conversely, to new forms of dispossession). In my case-study, cultural practices (or rather a mimesis of cultural practices) are also turned into sources of military value.

43. These works have been an enormous inspiration to me, as they make space for the subjectivities of Iraqis who suffered during these wars, alongside their analyses of occupation and war. See Zaynab Saleh's *Return to Ruin: Iraqi Narra-*

tives of Exile and Nostalgia (2021), Zahra Ali's *Women an Gender in Iraq: between Nation-building and Fragmentation* (2018), Omar Dewachi's *Ungovernable Life: Mandatory Medicine and Statecraft in Iraq* (2017), and Nadje Sadeg Al-Ali's *Iraqi Women: Untold Stories from 1948 to the Present* (2007).

44. Karl Marx critiques the transformation of man into a cog in a capitalistic apparatus, exposing the ways in which estrangement from one's own labor diminished human capacity to act as an agent in the world. In *Capital,* Marx notes that in the case of handicrafts and manufacture, "the workman makes use of a tool," whereas in the modern industrial factory "the machine makes use of him." That is, man become a "mere living appendage" of the factory ([1867] 2007, 461). For Marx, in both instances of labor, man is made part of a machine—turned into a living mechanism; however, in the alienated modern labor of the factory system, man becomes a grotesque appendage of that machine. Meanwhile, as man's labor power is objectified and he becomes increasingly estranged from it within that system, relationships between people begin to take on the character of things. Marx explains how this process occurs through his development of the concept of the commodity fetish, noting that commodities were "social hieroglyphs" that obscured the social relationships they embodied: "the commodity is therefore a mysterious thing because in it the social character of men's labor appears to them as an objective character stamped upon the product of that labor" (83).

45. Through the processes of mystification that Marx describes, "a relation between people takes on the character of a thing and thus acquires a 'phantom objectivity,' an autonomy that seems so strictly rational and all-embracing as to conceal every trace of its fundamental nature: the relation between people" (Lukács 1971, 83). Lukács emphasizes that the thingification of the relationships permeates the modern world as a whole, becoming imperceptible to its participants, appearing "as the only possible world" (110). For Marx and these thinkers who took up his concept, such reification extends from the world of economic exchange into human consciousness itself. Within this late capitalist hour in America's military-industrial complex, cultural trainings turn social relationships into highly reified, useable, and operationally relevant wartime modes.

46. Human technology might also be conceptualized as the tension between the gift relation and the commodity relation, or disentangled versus entangled life. To use a human being as a technology (and thus as a kind of thing) disarticulates that person from their web of relationships, essentially turning the gift of social relationships into a commodity. See Strathern (1988) and Tsing (2015). Thank you also to Isaiah Wilner Loredo for his thoughts on this matter.

47. Martin Heidegger traces the etymology of the word *technology* back to the Greek "*technikon* . . . that which belongs to *techne*," explaining that "what is decisive in techne does not lie at all in making and manipulating nor in the using of means" ([1954] 1977, 13). Conversely, he explains: "Technology is a way of revealing. If we give heed to this, then another whole realm for the essence of

technology will open itself up to us. It is the realm of revealing, i.e., of truth" (12). He offers, as explanation, the technology of an old wooden bridge, as it described in Holderlin's poem "The Rhine," wherein the thing itself—the river—remains a river, rather than a source of energy. The world is harmonious with and not effaced by the structure. However, for Heidegger, in the realm of "modern, industrial technology" a wholly different process ensues. Through the process of "Enframing," man turns the energies of nature into a "standing-reserve," that is, an arena that exists exclusively to be ordered and used. He then queries: "If man is challenged, ordered, to do this then does not man himself belong even more originally than nature within this standing-reserve?" (18). Although Heidegger initially balks at the idea of man himself becoming a useable resource in this manner (suggesting that man's position at the apex undermines this idea), he later cautions that it may indeed be possible for man to be transformed into a standing-reserve. For Heidegger, when man becomes, in his intrinsic identity, the orderer of the standing-reserve, he is on the verge of turning himself into a resource—in facing that abyss, all the more, he attempts to elevate himself further over that which he orders.

48. Much recent literature has focused on the increasing technicizing of warfare, culminating in posthuman warfare, where surveillance and targeting are outsourced to machines. At the beginning of this trend, Paul Virilio described how war had become increasingly prosthetic and mediated in the twentieth century (1989). More recently, P. W. Singer described what was framed in military circles as "the greatest revolution in military affairs since the atom bomb: the dawn of robotic warfare" (2009), and James Der Derian described the dawn of an ethically ambiguous "virtuous war," while Hugh Gusterson (2016) points to "ethical slippage" over time as drones are increasingly used. These works, along with a spectrum of academic and popular books, described how unmanned aerial vehicles were increasingly deployed by the American military with increasing sophistication, from wasp-sized surveillance pods to, more recently, the mass 3D printing of cheap, expendable drones (i.e., M. Benjamin 2012; B. Williams 2013; Bergen and Rothenberg 2014; Chamayou 2015). Other recent scholarship has examined the intersection between a posthuman analytic and nonhuman animals, examining how other species help to constitute war's possibilities (i.e., Cudworth and Hobden 2015).

49. Department of Defense June 2007 Summit, "Regional and Cultural Capabilities: The Way Ahead, Regional and Cultural Expertise."

50. Indeed, in a much-circulated *New York Times* article in 2015, Joseph Goldstein describes U.S. military tolerance of the rape of Afghan boys by Afghan police officers, as they were instructed to treat it as a cultural practice.

51. Leila Ahmed points to how Muslim women have been constructed as oppressed by colonial rule, and consequently how Muslim and Middle Eastern culture is read as patriarchal and timeless. Lila Abu Lughod has pointed out that these deflections into the woman question and away from geopolitics and power

have become justifications for violent interventions. Zahra Ali describes the "gender dimension" of cultural imperialism.

52. In "The Age of the World Target: Atomic Bombs, Alterity, Area Studies," Rey Chow (2010) posits that "scientific" and "objective" knowledge production about "areas" of the world during peacetime work to help constitute the world itself as a target—and that this knowledge production perhaps "shares the same scientific and military premises as war," proposing that such attempts to "know" another culture are "doomed to fail." She writes: "Can 'knowledge' that is derived from the same kinds of bases as war put an end to the violence of warfare, or is such knowledge not simply warfare's accomplice, destined to destroy rather than preserve the forms of lives at which it aims its focus?" (16).

53. While the Green Berets participated in cultural identification exercises, they also have more complex trainings (i.e., Unconventional Warfare). I refer to these exercises specifically when I analyze them.

54. The contours of these trainings are elaborated further in chaps. 1–2 and their footnotes.

55. In *Simming: Participatory Performance and the Making of Meaning* (2014), Scott Magelssen describes cultural simulations enacted at Fort Irwin, the Army's National Training Center. In Anna J. Simons's *The Company They Keep: Life Inside the U.S. Army Special Forces* (1997) and in Dick Couch's *Chosen Soldier: The Making of a Special Forces Warrior* (2008), both authors describe the unconventional warfare exercises of the Green Berets, such as Robin Sage, and more broadly, the cultural simulations that are part of the trainings.

CHAPTER 1. THE MAKING OF HUMAN TECHNOLOGY

1. It is essential to make a distinction between the architects of military policy and those who enact it on the ground.. Military designs are not the work of soldiers, but of power brokers. Furthermore, those designs are enacted on the ground variably and imperfectly.

2. In her essay, "Culture as a Weapon System," Rochelle Davis describes the various reifications of U.S. military categorizations of Iraqis and how they changed over time, from more amorphous assertions about "Iraqiness" (focusing on valorization of family and religion) to an orderly taxonomy of categories of Iraqi difference, in particular emphasizing primordial sectarian antipathies. The earliest cultural materials authored by and for the U.S. military "described Iraqi culture with recourse to the national character studies that typified the culture research and cultural anthropology of the 1940s and 1950s" (2010, 9). Davis explains that in the first U.S. military handbook to Iraq published in 2003, Iraqiness was described as a timeless essence, "uniquely determined by family and religion," rather than as "a product of history or political forces or government

policies" (2010, 9), hearkening back to Ruth Benedict's war ethnographies of Japan (1946) and their emphasis on cultures as patterned wholes. Following this idiom, the first U.S. military handbook to Iraq described Iraqiness as its own monolithic pattern, which delineated a reified "Arab worldview" constituted by characteristics such as "deep belief in God"; "a desire for modernity contradicted by a desire for tradition"; "paranoia" and privileging family over self (Davis 2010, 12). However, only a year later in 2004, Iraqi culture was re-represented by the U.S. military as primordially riven by sectarian division, significantly before the fighting in Iraq became sectarian in 2005 and 2006.

3. The unit of the *tribe*, which received particular attention from the American military in the early years of the war, offers a brief and telling case study. David Kilcullen, senior counterinsurgency advisor to General Petraeus in Iraq, proposed that the category of the tribe trumped all other identifications: "When all involved are Muslim, kinship trumps religion: the key identity drivers are tribal" (in Gregory 2008, 14). According to the *U.S. Counterinsurgency Field Manual 3-24*, tribe constituted the primary form of social identity in Iraq, and was thus the key to undermining the insurgency: "essentially, diminishing support for the insurgency entailed gaining and maintaining the support of the tribes" (2006, 4–7).

This assumption undergirded, for example, the U.S. military's creation of an Iraqi tribal militia called al-Sahwa (the Awakening Council) to counter what seemed to be less knowable insurgent forces. In this calculus, the tribal identification, imagined as the most powerful and most primordial social form in Iraq, would be animated within a structure useful to the U.S. military. However, the U.S. military had reified and obscured a fluctuating social form, turning it into a social category, adopting previous colonial reifications. Hundreds, if not thousands of members of al-Sahwa left their tribal coalitions, some of whom defected to al-Qa'ida: a confirmation of how forms of social difference are not static but are rather deployed historically and politically.

Similar reifications occurred with devastating consequences around the term *sect* in the wake of the imposition of American military categories. Rather than an eternal or immutable category, the Iraqi population's relationship to sect shifted over time, amidst the Iran-Iraq War, the sanctions thereafter, and culminating in the American occupation in 2003. Fanar Haddad describes the shift from "banal sectarianism" (a simple awareness of sect) to "aggressive sectarianism" (2011). Peter Sluglett notes: "U.S. insistence that sectarianism is one of the organizing principles of Iraqi society, and even more, that it forms the basis of Iraqi politics, has done incalculable harm" (2015, 259).

4. Ethnographers and theorists caution conversely that ethnic boundaries and tribal identity are socially and historically produced and a product of social organization (Barth [1969] 1998; Eickelman [1981] 2001). Indeed, the tribe in Iraq was a profoundly mutable form of social difference that morphed into an objectified category only through colonial discourses and policies.

In his historiography, Hanna Batatu (1978) describes the fluctuating position of the tribe, as it was subject to the Empire's reconfigurations. For example, although the Ottoman period in Iraq (1534–1920) was largely categorized by discordant and fragmented tribes with only tenuous links to the cities, the Empire sought to Ottomanize its provinces between 1831 and 1914, and to reconfigure tribal loyalties through the medium of land. The Ottoman land law of 1858 designated all land (apart from *mulk,* absolute private property, and *waqf,* land allocated for Islamic purposes) as *mīrī* owned by the state and required the registration of land ownership.

This policy prompted the creation of numerous small tribesmen landowners, the transformation of shaykhs from recipients of tributes to tax-farmers, and the fostering of rivalries between beneficiary and nonbeneficiary tribes. To this end, the political and military content of the shaykh's position was dramatically diluted, while his new individual economic power became entrenched in "new subversive relations of production," and dependency on the state. Hierarchies that had previously depended on a network of patriarchal blood-relations were supplanted by more fixed property relations, as the Ottoman Empire became increasingly implicated within the industrial capitalist world system. Meanwhile, the previous tribal system was left in fragmentation and disarray: "In Baghdad Wilayat alone . . . there were in 1918 at least 110 independent tribes, made up of 1,186 sections. Many of these sections were practically free from the authority of the parent tribe" (Batatu 1978, 77).

5. For example, the period of the British Mandate in Iraq can be read as yet another radical reconfiguration of the "tribe," and its transformation into a primary and original category of difference rather than a malleable form. Batatu asserts that in a "reversal of history," the British *artificially* reanimated a tribal system that was dramatically declining in order to create political equilibrium: "in the balancing of the tribesmen with the townsmen, [they saw] the surest guarantee of the continuation of their own power" (1978, 24). To this end, the British created a shaykhly class-for-themselves: landowners dependent on the Mandate for their privilege.

The British proclaimed that tribal custom had the force of law and excluded the tribal countryside from the purview of the Iraqi Constitution of 1925, as the shaykhs were transformed into an anchor for the British and medium of administration of the countryside amidst diminishing colonial resources. Beyond the question of expediency, the figure of the tribe took on a romantic and primordial valence in British imagination, in part a foil to the urban Ottoman effendi, represented as frequently corrupt and despotic (Dodge 2003, 61).

According to Toby Dodge in *Inventing Iraq: The Failure of Nation Building and a History Denied,* it was the British notion that Iraq was "pre-modern and 'rural,' untainted by the negative and destabilizing effects of capitalism" that made the shaykh and his tribe "'naturally' the dominant institutions through which British policy aims were to be realized" (2003, 83). Shaykhly authority, imagined as natural and consensual—and not perceived as being actually on the decline—was thus mapped and instrumentalized by the British.

According to Dodge, the British thereby "inadvertently but radically changed the nature of the shaykh's actual relation to the rest of Iraqi society" (2003, 83). Indeed, Dodge further notes the startling parallels between the British colonial categories of social difference that were instantiated in Iraq and echoed by the American military.

6. See Zainab Saleh (2021) regarding opposition Iraqis in exile in London. Certain Iraqis in exile with opposition politics, such as Fouad Ajami, Kanan Makiya, and Ahmed Chalabi, had particular influence on U.S. policy in Iraq. In the run-up to the war, the State Department put together a series of committees of Iraqi exiles to discuss the future of Iraq. These collaborations and a commitment to organizing Iraq into a federalist structure rendered sectarianism a key lens. See also Hamid Dabashi (2011) on the comprador function of the "native informant," and discussions specifically of Ajami and Makiya. Dabashi positions Azar Nafisi's *Lolita in Tehran* as the native addendum to works like Fukuyama and Huntington on civilizational conflict: that is, the native informant is also part of legitimizing Empire's strategy of domination.

7. Critics of the so-called Cultural Turn have pointed to what Ann Stoler calls a focus on the "intimate" in the consolidation of power, particularly in colonial and neocolonial contexts (2002). The DOD's Minerva Initiative and the Army's Human Terrain System have indeed entailed the diagramming of the social structures and the harnessing of information about potential adversaries' and allies' social patterns as well as motivations: the sphere of Stoler's intimate. As Stoler asserts, such classification is "not a benign cultural act but a potent political one" (2002, 8). See also Marilyn Ivy's critique of Benedict, wherein transforming cultures into patterned wholes generated a contrapuntal effect, a self and other bound in opposition: "Japan became America's evil twin. . . . Likewise, America became Japan's good twin" (2008).

8. As described in "Building Language Skills and Cultural Competences in the Military: DOD's Challenge in Today's Educational Environment," published by the U.S. House of Representatives Subcomittee on Oversight and Investigations in 2008: "language" and "regional training" were increasingly regarded as "mission critical skills." "Foreign language and regional proficiency . . . shall be considered critical to the continuum of professional military education and training" (2007). After the 2005 launch of the Defense Language Transformation Initiative, focus on language and regional skills escalated.

9. In chaps. 1 and 2, I discuss the trainings that were available to the conventional forces during the height of the 2003–2011 Iraq War and were the backbone of the emergent Cultural Turn. Those trainings became more robust over the years, as documented by the "Army Regulation 350-1: Army Training and

Leader Development" (Department of the Army 2014), which outlines the variable range of trainings, dependent on deployment and mission.

At present, all deploying units must complete the Defense Language Institute Foreign Language Center (DFLIC)'s "Rapport Program" a "four to six hour DL [Defense Language] tool that introduces the soldier to the language and culture of a specific contingency theater. Every deploying soldier trains with Rapport, however, one soldier per deploying platoon undergoes a more rigorous curriculum of culturally based language training either in the form of a 16 week language training detachment or completion of 'Head Start 2,' an 80 to 100 hour DL platform available in 12 strategic languages. All Soldiers must also be able to use the Language Survival Kit" (2014, 143). By contrast, Green Berets have always had significantly more robust language and culture trainings. All trainees receive eighteen to twenty-four weeks of language and culture training, depending on language difficulty, e.g., eighteen weeks for French and twenty-four weeks for Arabic (Special Forces Association).

10. Post 9-11 political logics were indebted to a technophilic Cold War politics and optics of rationalized planetary oversight within a chaotic world (see, in particular, Edwards 1996). These logics of omniscience have culminated most recently in an increasingly mediated military technoscape, where Paul Virilio elaborates a twenty-first-century military fantasy and teleology of "ubiquitous, orbital vision of enemy territory" (1989, 2). Amidst the increasing use of techniques of cinema in the evolution of war technologies (from the watchtower and remote balloon to the camera-equipped reconnaissance aircraft, culminating in the contemporary Predator drone), Virilio cautions that progressive-seeming magnification is coupled with distortion. As the "image was starting to gain sway over the object [and] time over space" (1), there is simultaneously a profound "derangement of perception" wherein target locations increasingly resemble cinema locations (72).

11. In his monograph on national security affect, Joseph Masco notes that: "the failure to predict global events . . . haunts U.S. security culture today, creating a constant drive for new technical capacities and increasing militarization of American life" (2014, 6). Masco locates the American response to that anxiety within an affective circuit: "The ability to shock (at both psychic and material levels) and not experience shock . . . . became the primary goal of the American security state after 1945" (9). Masco proposes that to counteract these anxieties, the American security state consequently expands to a planetary scale.

Within a similar logic, Paul Virilio has delineated an American military fantasy of increasing posthuman omniscience via prosthetic amendments culminating in the Iraq War of 2003 (2009). He points to a progressive conjoining of the weapon and eye and the codevelopment of war and cinema technologies since the Industrial Revolution, as he articulates this military teleology. Rey Chow, meanwhile, links knowledge production and Area Studies as a means of turning the world into a target field (2010, 16).

12. In the aftermath of the Cold War and then 9-11, the American work of war has inhabited an ongoing contest between presence and absence. The unipolar logics of the Cold War were increasingly supplanted by a military vision of adversaries that were more dispersed among the population and thus more illegible, prompting a focus on engaged presence of soldiers on the ground to counter these irregular threats. Between 2001 and 2011, over 2 million American soldiers were deployed to Iraq or Afghanistan, compared to the 3.5 million who served in the Southeast Asia theater. Though casualties in Iraq and Afghanistan were less grave *by tenfold* than in the Vietnam War, the military deployment numbers in the post 9-11 wars have been considerable. (In the Vietnam War, approximately 60,000 American soldiers were killed, compared to approximately 6,000 in Iraq and Afghanistan, according to the U.S. Military Casualty Statistics published in 2013.)

13. Incipient military logics emphasized the importance of soldiers negotiating the intimate complexities of urban warfare with their "boots on the ground." Simultaneously, however, there was an emergent trend towards the opposite: the outsourcing of certain forms of wartime action to others. After the fall of the Berlin Wall, private markets rushed to fill in a gap in security, prompting the increasing privatization and corporatization of war. Moreover, the Iraq War brought an unprecedented explosion of outsourcing of military and government functions that occurred during the 2003 Iraq War.

P. W. Singer describes how private markets rushed to fill a "security gap" (2003). Naomi Klein notes that in 2003, the U.S. government handed out 3,512 contracts to security companies to protect the United States. By 2006, there were 115,000 such contracts (2007). In this period, Privatized Military Firms (PMF) took over security functions abroad, minimizing transparency and the numbers of military deployments from the state, while Unmanned Aerial Vehicles combed the streets for targets without threat to the soldier's life. Amidst this confluence of trends—the imperative of the soldier to be on the ground, while in other ways dislocating from that ground—the U.S. Department of Defense sought to navigate these contradictions: the local intermediary was the means of lifting U.S. boots slightly off the ground.

14. For more on PMESII-PT, see, for example, *Army Doctrine Publications 3-0 Land Operations*, www.army.mil/e2/downloads/rv7/info/references /ADP_3-0_ULO_Oct_2011_APD.pdf.

15. In *Orientalism*, Edward Said dismantles and exposes colonial discourses of a charged, covert, and illogical "inside"—the harem, the "Oriental" market, and the village. Within these idioms, the Orient is an entity to be penetrated: the "interior" might be the capturable commercial resources of a colony, or more insidiously the knowability and co-optability of the colonized themselves. Said writes: "The point here is that the space of the weaker or underdeveloped regions like the Orient was viewed as something inviting French interest, penetration,

insemination—in short, colonization" (1978, 219). Edward Said's notion of the colonial inside offers as a charged and ambivalent space that nonetheless might be domesticated and "known." Additionally, Freudian notions of the interior as a vast unknown, a "dark continent" accessible only via psychoanalysis, is of course pivotal in this intellectual genealogy. In my analysis, I look to conceptions of the "inside" and "interior" as they pertain to Empire. Indeed, Ranjana Khanna makes parallels between Freud's description of female sexuality as a "dark continent" and colonial metaphors for Africa (2003). One crucial precursor here is Joseph Conrad's notion of the "interior" as accessed through progressive penetration into land, into war, and into the underbelly of "civilization" in tandem with an inward journey into the self (1899).

16. Edward Said quotes Maurice Barrès on the importance of a cultivated class of native insiders (in this facilitating an insider for a colonizer), "who would thus form a link between us and the mass of natives" (1978, 245). In his sociology of knowledge, Robert Merton describes an imagined epistemological bifurcation between "insiders" and "outsiders," which considers the "Insider as Insighter, one endowed with special insight necessarily obscure to others"—or, as he boils it down, "you have to be one in order to understand one" (1972, 15). In this regard, the deployment of intimacy, trust and rapport as tools are means of entry into the "inside" (Said 1978). This notion of becoming an insider is perhaps somewhat beyond Clifford Geertz's notion of "finding one's feet" in the field (1973). Rather, the notion is more explicitly bound up with Empire, as the soldier is asked to become something like a cultural insider to further military aims.

17. According to an operations officer from the U.S. Army and Marine Corps Counterinsurgency Center: "T. E. Lawrence has in some ways become the patron saint of the U.S. Army advisory effort in Afghanistan and Iraq." Moreover, during U.S. campaigns in Iraq and Afghanistan, military theorists have resurrected Lawrence. John Nagl, a coauthor of the *Counterinsurgency Field Manual* (FM 3-24), noted the importance of Lawrence in an interview: "What Lawrence gave us was an appreciation of how difficult the task is in understanding that we have to work through our local allies. And also a sense of the independence of thought and action that is required both for good insurgent leaders and, in many cases, for good counterinsurgency leaders" (Soussi 2010). A Green Beret in his thirties named Jack once told me: "Lawrence is beyond doctrine. He is canon." On another occasion, Fred, a Vietnam-era Green Beret explained: "We've had many classes about Lawrence of Arabia. He's taught as a model. But he's not just taught in [Special Forces], he's taught in War College. He's the model for how to work with indigenous forces." Indeed, at the outset of the U.S. invasion of Iraq, *Seven Pillars of Wisdom* was number two on a recommended reading list for officers that was compiled in the "Inside the Pentagon" newsletter.

18. The focus on lightness and even imperceptibility—indeed to almost vanish *into* another cultural terrain (at least among the Special Forces) ripples

backwards into the existential genealogy of the discipline of anthropology itself and its early colonial imbrications. In 1825, British Orientalist Edward William Lane journeyed to the Ottoman Empire, went to Cairo, and disguised himself as a Turk for two and a half years to acquire the trust of the Turks and thereby acquire more nuanced insider knowledge. Fifty years later, Frank Hamilton Cushing, having been assigned by the Smithsonian Institution's Bureau of Ethnology to collect information about the Zuni, learned to speak fluent Zuni and was inducted into both a Zuni Pueblo and the Bow Priesthood. Hamilton Cushing wrote euphorically about the lure of going native. Having already spent two years among the Zuni, he wrote: "I would be willing to devote, say, a year or two more to it to study for a period almost as great from the *inside* the life of the Zuni, as I have from the outside" ([1879] 2011, 6). His contemporaries too courted this kind of entry. For example, anthropologists Walter Spencer and Francis Gillen claimed initiation into the tribe they studied, emphasizing the necessity for comprehension of entry into "the mental attitude of the savage" ([1899] 2013, 54).

19. The complex entwinements between sentiment and power within imperial contexts have received increasing ethnographic attention. Considering how "sentiment" is used as a tactic of governance, Stoler asks: "What sorts of self-mastery and sensibility were cultivated among civil servants, which empathies were enlisted?" (2009, 59). Danilyn Rutherford considers the charged effects of "colonial sympathy" (2009) while Audra Simpson (2014) interrogates the entwinements between sovereignty and sympathy. Didier Fassin (2012), meanwhile, examines how moral sentiments are deployed within contemporary politics of humanitarianism, and Carole McGranahan (2018) examines experiences of "love" within these contexts. See, in particular, Carole McGranahan and John Collins, eds., *Ethnographies of U.S. Empire* (2018).

20. Popular media and public discourses have frequently lauded empathy as a balm for all social ills, as a bulwark against war, and as a means of holding society together (Hoffman 2001; Howe 2012; Krznaric 2015). Conversely, empathy and its variable embodiments and consequences are best put in context without a single value judgment (Holland and Throop 2011; Skultans 2008). In my article, "Imperial Mimesis" (2018), I describe how when "empathy" (i.e., emotional connection and identification) is instrumentalized as a weapon, the effect is typically either mutual estrangement (what soldiers call "going cold") or excessive intimacy ("going native"). Meanwhile, role-players experience many of these moments of interface as puzzling and alienating.

21. In "Imperial Mimesis," (2018), I describe how soldiers are trained in a mimetic form (miming the outward gestures and postures of their interlocutors) that is largely devoid of substance (truly empathizing with their interlocutors' lived experience). In imperial mimesis, the soldier tries to make the distant proximate by enacting an image of the Other (through gesture, posture, etc.) in order to gain power over those whom they mime ("locals" and, ultimately, potential

terrorists). In an inversion of classic mimesis, a mechanism of subversion instead becomes a means of dominance.

22. For many soldiers I spoke to, cultural sensitivity coupled with the Rules of Engagement made soldiers feel that they couldn't protect themselves.

23. Rutherford's notion of colonial sympathy overlaps with Chinua Achebe (1988, 149), as discussed by Robert Foster. He proposes that such sympathies create an "imaginative identification"—namely, "a self-encounter that enables us 'to re-create in ourselves the thoughts that must go on in the minds of others, especially those we dispossess'" (Foster 2001, 66, in Rutherford 2009, 5).

24. Gant, described by some military personnel as "Lawrence of Afghanistan," for having "gone native," was later demoted then forced to retire, after having an affair with a reporter and violating other military regulations. Soldiers I spoke to alternately viewed Gant as a cult hero and as a controversial figure. https://time .com/2921469/the-fall-of-the-green-berets-lawrence-of-afghanistan.

25. "Close with and destroy" is the training mantra of infantrymen in the conventional army.

26. Barack Obama, "Sustaining U.S. Global Leadership: Priorities for 21st Century Defense," 2012.

27. Indeed, in 2017, President Trump dropped 9,000 more bombs on ISIS than President Obama had in 2016 (40,000 to Obama's 31,000). Meanwhile, he increased American support for Saudi Arabia's war in Yemen, more than doubled troop levels in Somalia, sent more than 3,000 troops to Afghanistan, and authorized an indefinite U.S. military presence in Syria. Meanwhile, although President Obama put in place reforms to make the drone program more transparent, the Trump administration dismantled these safeguards, expanding the sites drones could be used and not publishing civilian casualty data. In 2019, the U.S. deployment figures were up to 200,000, in Afghanistan, Syria, Iraq, Saudi Arabia, and the Gulf—more troops than when Trump came into office.

See www.vox.com/world/2018/1/30/16925544/trump-state-union-2018-military-2017; www.foreignaffairs.com/articles/2019-12-03/trump-didnt-shrink-us-military-commitments-abroad-he-expanded-them; www .commondreams.org/views/2020/01/13/droning-world-assassination-complex-bush-obama-trump;www.nydailynews.com/news/politics/ny-pol-trump-revokes-rule-intel-drone-deaths-20190306-story.html; www.nytimes.com/2019/10/21/world /middleeast/us-troops-deployments.html.

28. Nick Turse, "Donald Trump's First Year Set a Record for Use of Special Operations Forces," *The Nation,* Dec. 14, 2017, www.thenation.com/article /donald-trumps-first-year-set-a-record-for-use-of-special-operations-forces.

29. See President Biden's approval of the "Global Posture Review," and additional coverage in www.defense.gov/News/News-Stories/Article /Article/2856053/biden-approves-global-posture-review-recommendations; www

.nytimes.com/2021/11/29/us/politics/biden-us-troops-review.html; https://
theintercept.com/2021/12/02/biden-military-deployment-global-footprint.

30. David Vine points to lily-pad bases in *Islands of Shame* (2009).

CHAPTER 2. THE IRAQ WARSCAPE AND
THE CULTURAL TURN

1. In this chapter, I focus largely on the experiences of General Purpose Forces
rather than the Special Forces. It was with the former group of soldiers that the
Cultural Turn took effect most visibly as an institutional shift and for whom
many of the present Iraqi role-players formerly worked as interpreters and
contractors.

2. For the Iraqi death toll, see the Watson Institute, "Costs of War," https://
watson.brown.edu/costsofwar/costs/human/civilians/iraqi; for the estimates of
displaced Iraqis, see UNHCR, " Iraq Refugee Crisis," www.unrefugees.org
/emergencies/iraq.

3. According to Emily Apter, war might be understood as "a condition of nontranslatability or translation failure at its most violent peak" (Apter 2006, 19).

4. The analysis of translation has a long and rich theoretical genealogy with a
wide range of concerns, including questions of fidelity to origin texts and afterlives of texts; translatability or untranslatability; amplification and renewal versus diminishment and loss (i.e., Benjamin [1968] 2007; Derrida 1978, 1985a,
1985b; De Man 1986). In "The Task of the Translator," Walter Benjamin posited
that the work of translation in its ideal form created a vital afterlife, extending
the original text, critiquing prior notions that translation should reproduce the
origin text identically: "it is translation which catches fire on the eternal life of
the works and the perpetual renewal of language"([1968] 2007, 74), inspiring
extended poststructuralist conversations on the extent to which all representation and meaning-making constituted acts of translation.

5. In the 1990s, a cultural turn arrived in translation studies, such that "neither the word, nor the text, but the culture becomes the operational 'unit' of translation studies" (Bassnett and Lefevere 1990, 8), leading to multicultural and postcolonial lenses. Postcolonial thinkers of translation have urged in particular the
necessity of locating the movement of texts within complex power relations and
hierarchies (Spivak 1988; Ashcroft, Griffths, and Tiffin [1989] 2003; Bhabha
1994; Bassnett and Trivedi 1999). For example, in *The Empire Writes Back*, Ashcroft, Griffiths and Tiffin contend: "The standard problem—'is translation really
interpretation'—familiar to all translation theory, becomes magnified in the postcolonial setting. Who translates whom becomes a crucial issue" (204). Indeed,
translation has a long history of entanglement with power. Vincente Rafael quotes
the Spanish humanist Antonio de Nebrija, in 1492, who notes that: "Language is

the perfect instrument of empire reminding us that all imperial projects through-out history have relied on the work of translation and translatability (1993, 23). Talal Asad has since critiqued cultural translation's enmeshment in power.

6. Colla critiques the alarming intersection between the humanist narrative of translation as a means of cultural understanding and hope, and a military narrative where translation is a means of "bridging civilizations." Pointing in particular to the post 9-11 ballooning of the Arabic-English translation industry, Colla us reminds that fantasies of "cultural bridging" are entrenched in power asymmetries (2015, 4).

7. In her ethnography of Palestinian soldiers in the Israeli military, *Surrounded* (2008), Rhoda Kanaaneh describes the contradictory challenges and hence strategies that have been adopted, which does not mean a lack of allegiances. She suggests that human beings maximize their possibilities amidst constraint in times of war and occupation: "Humans do not make free choices. Depending on their perceptions of the opportunities before them, they can put their eggs in different baskets" (16). Further, people mobilize and demobilize different identities strategically (in her case study, Palestinianness, Arabness, and Israeliness). She explains: "Their mobilizations are informed by the 'basket' at hand—not being harassed at the mall, 'getting somewhere' in this region, or wanting a piece of the Oslo pie—the parameters of which are often set by the state" (25). I will discuss *sect* in depth later in the chapter. See also Kali Rubaii's (2019) discussions of how sectarianism became a "social fact" through the continuous performance of a convergence between sect and place.

8. Haydar Al-Mohammad urges describing Iraqi lives as caught not merely in the forms and structures of tribal obligations and sectarianism, and the violence and destruction of terror, but also in the rough ground of mundane affairs and encounters, and its incorporeal ethics. This "rough ground" relies on an ethics of embeddedness (developed from Wittgensteinian embeddedness of language in its contexts of use). I follow Al-Mohammad in this commitment. I add adjacent the necessity of performances of tribal obligations and sectarianism within certain charged spaces of affiliation (as in Iraqis who were demonstrating their commitments to Americans or other Iraqis).

9. See Dewachi (2017) for an extensive analysis of how British described Iraq in similar terms: wherein people, land, and rivers, are all narrated as unknowable, unnavigable, illegible, illusive, and so on.

10. See Zainab Saleh (2020), in particular, for discussions of Iraqis in exile in London and the groups that congregated there, including variegated opposition to Saddam Hussein's regime. For example, the Iraqi National Congress (INC) under Ahmed Chalabi and the Iraqi National Accord (INA) under Iyad Allawi emerged in the 1990s, and would both become key players in post-Saddam Iraq. Chalabi in particular fostered close relations with Neoconservatives in the United States, who were already pressuring the U.S. government to remove

Saddam Hussein from power. With the election of President George W. Bush, Iraq became a central policy focus after September 11, especially after the U.N. did not uphold sanctions or continue weapons inspections of the country.

11. The position of the Iraqi population towards the war was complex and should not be flattened, encompassing everything from pro-"liberation" opposition returning from exile (see note 10) to Iraqi intellectuals who perceived the American presence as an occupation from the beginning of the war, bringing the moment into historical conversation with nationalism, imperialism, and everything in between. For more regarding the opposition of the intellectuals, see Dr. Saad Jawad, a political scientist at Baghdad University; a consolidated intellectual movement of resistance was slower to form because of tensions among intellectuals regarding the hatred of Saddam Hussein's regime and a desire to build a new Iraq, and subcommunities of intellectuals that formed specifically along sectarian lines (Jawad 2006). See also Muhsin al-Musawi's *Reading Iraq: Culture and Power in Conflict* (2006) for more on the position of Iraqi intellectuals.

12. In a poll conducted in 2004, 82 percent of Iraqis said they disapproved of the United States and allied militaries in Iraq. See Thomas E. Ricks, *Washington Post,* May 13, 2004, www.globalpolicy.org/component/content/article/166-advocacy/31145.html.

13. Thiranagama and Kelly note that political betrayal is often a product of "contradictory social and political obligations," where individuals negotiate competing and "fundamentally incomplete" loyalties (2010, 1–2). Still, political life, they contend, often takes place in gray zones, "destabilizing the rigid moral binaries of victim and persecutor, friend and enemy" (1). Accusation is starker, naming who is inside and outside and thereby "laying claim to political and moral certainty in the face of uncertainty" (2). Likewise resonant with the Iraq case study, they note that the modern state itself is "built on betrayal"—as prior allegiances are replaced with new ones; the Iraqi state as such was not viewed as legitimate under Saddam Hussein and consequently forming new allegiances was not in and of itself viewed as treasonous by many Iraqis.

14. Elliot Colla (2015, 135) explicitly describes these individuals as "conscripts of modernity," in Talal Asad and David Scott's sense. As Asad notes, amidst shifting power configurations, people inhabit and make choices within conditions not of their own making: "For conscripts of Western civilization this transformation implies that some desires have been forcibly eliminated—even violently—and others put in their place" (Asad 1992, 345). Scott concludes that amidst shifting postcolonial realities, new subjectivities are constructed in altered relations of power and subaltern resistance might not be the most relevant paradigm. Rather, "what is at stake here is not whether the colonized accommodated or resisted but how colonial power transformed the ground on which accommodation or resistance was possible in the first place" (Scott 2004,

119). As these individuals made the choice to work with the American occupying force, they were operating within conditions not of their choosing.

15. See, for example, Hamid Dabashi (2011) on his category of "native inform-ant" and the comprador intellectual specifically, as upwardly mobile intellectu-als who legitimized Empire's position. He specifically excludes here the transla-tors of GWOT, focusing instead on a different category of elite (he names as examples Irshad Manji, Ayaan Ali Hirsi, and, in the case of Iraq specifically, Kanan Makiya, Fouad Ajami, and Ahmed Chalabi).

16. Available texts on Saddam Hussein's regime are scant and politically charged, as they are typically linked to diasporic Iraqis who assisted the Bush Administration in the run-up to the invasion of Iraq. I thus supplement these problematic texts with ethnographic data.

17. According to Aburish, "An unknown number of innocent people fell victim to the pettiness of personal vendettas with no political content."

18. In Feldman's study of Gaza under Egyptian rule, public participation in policing was encouraged as part of a security strategy. Unable to be everywhere at once, the Egyptian security forces generated a collective hermeneutics of suspicion and participation. Fear and uncertainty were converted into practices of "inform-ing, self-policing, and control of others in their community" (Feldman 2015, 30).

19. In Verdery's study of Romania's Transylvania region in 1973, during the rule of the communist dictator Nicolae Ceauscescu's Securitate, the secret police and the public's role in reporting created "a constant current of mistrust and doubt [that] eats away at trusting relations" (Verdery 2018, 19) as people were both spied on and assumed themselves to be spies.

20. Arendt asserts: " . . . that the psyche *can* be destroyed even without the destruction of the physical man; that indeed, psyche, character, and individuality seem under certain circumstances to express themselves only through the rapid-ity or slowness with which they disintegrate" (1976, 441). In Arendt's reading, in totalitarian settings, the human interior is invariably eventually co-opted.

21. In *Ambiguities of Domination: Politics, Rhetoric, and Symbols in Contem-porary Syria*, Wedeen's case study on Hafiz al-Asad's regime, she concludes that Asad's spectacular cult "operates as a disciplinary device, generating a politics of public dissimulation," a pretended reverence that leaves the interior intact. She proposes that this politics of "as if" "provides guidelines for acceptable speech and behavior; it defines and generalizes a specific type of national membership; it occasions the enforcement of obedience; it induces complicity by creating prac-tices in which citizens are themselves 'accomplices,' upholding the norms consti-tutive of Asad's domination" (1999, 6). Wedeen draws upon Vaclav Havel's notion of the "social auto-totality," wherein citizens reinforce one another's compliance without believing in that system. Meanwhile, although Wedeen looks at exterior presentation in relation to a contrastive interior state, the constitution of subjec-tivity is not reducible to the public performance of obedience to the regime.

22. I look here at tropes of "performance" through the lens of Goffman rather than choosing a Butlerian logic of "performativity," as I do not claim here that these performances are necessarily or uniformly constitutive of these individuals as subjects, on the level of the psyche. Rather, these moments of performance are crafted as a vehicle of achieving desired ends. In his discussion of performance in everyday life, Erving Goffman defines "performance" as all the activity of a given participant on a given occasion, which serves to influence in any way any of the other participants (1959, 12). He describes daily life in dramaturgical as well as utilitarian terms, suggesting that an individual "mobilize[s] his activity so that it will convey an impression to others which it is in his interests to convey" (4). The performances he analyzes constitute intricate games, where individuals employ ongoing precautions to make sure their definitional claims are not interrupted, what he calls "defensive practices" (13).

23. In his case study on Palestine, Tobias Kelly muses: "The fear of the collaborator . . . is not simply that when you 'tear off the mask' there will be an Israeli underneath but something far more frightening: that just possibly, there will be nothing" (2010, 187).

24. See also Darryl Li (2019) about the presumed and intrinsic immorality associated with mercenaries (positioned as less moral than violence driven by nationalism). Thank you to Kali Rubaii for this insight.

25. In *Shock Doctrine*, Naomi Klein describes the intersections of war with opportunistic and unfettered capitalism. In the case of the 2003–2011 Iraq War, the U.S. chief envoy L. Paul Bremer launched major economic shifts in the country: "mass privatization, complete free trade, a 15 percent flat tax, a dramatically downsized government. Iraq's interim trade minister, Ali Abdul-Amir Allawi said . . . 'we don't need this shock therapy in the economy'" (9). Klein documents the Bush administration's dramatic outsourcing of key war functions in the Iraq War to for-profit companies "from providing health care to soldiers, to interrogating prisoners, to gathering and 'data mining' info on all of us" (12). She notes the scope of this shift: in 2003, at the beginning of the war, the American government gave 3,512 contracts to security companies; three years later, at the sectarian height of the Iraq War, there were no fewer than 115,000 contracts given to security companies. In this paradigm, there is a "closed profit-loop of destruction and reconstruction" (381): security firms make money during all stages of war. Klein works to debunk the myth that unregulated capitalism and democracy go hand in hand. She also points to the corruption of key officials in the Bush administration who brought governance and business interests together.

26. According to the Rand Corporation's report by Dave Baiocchi, "Measuring Army Deployments to Iraq and Afghanistan" (2013, 3), the U.S. provided nearly two million troop years to the two deployments. A troop-year is a metric used to measure cumulative deployment length. For example, one troop-year is equivalent to any of the following cases: one soldier spending twelve months deployed,

two soldiers deployed for six months (each), one soldier deployed for eight months and another soldier deployed for four months, or twelve soldiers deployed for one month each.

27. See also Noah Coburn's *Under Contract: The Invisible Workers of America's Global War* (2018) on the impacts of contracting labor in wartime, and the ambiguities in many of its processes (secrecy around application process, legal ambiguities, etc.) for individuals who sought to "balance financial benefits of contracting with risks of injury and exploitation" (70). Coburn describes a system that was itself often corrupt, nontransparent, and poorly managed, where companies and officials benefit and workers suffer, making it difficult to find data on contractors and requiring workers to rely on personal networks.

28. These observations are supported by my own interviews. Additionally, see also Barakat (2005); Cockburn (2007); Wehry (2010); Al-Mohammad (2012). Patrick Cockburn contends that it is likely that many kidnappings in this period (even those of Westerners) were not explicitly political, but that political, ideological, and religious slogans were appropriated to get more media attention. Regarding kidnapping as an industry in this period, see also Napoleoni (2005) and Al-Mohammad (2012).

29. The Al-Ma'āni Arabic-English dictionary indicates that the word means "to eat and drink." Medieval dictionaries such as Lisan Al-Arab conjured a richer spread of meanings: eating, especially wheat, and the tough kernels that were difficult to cleanse and were eaten in impoverished times; forms of embroidery, the repetitive adding of small detail around the edge; and forms of talk or verbal disclosure that also conjured repetition, like gossip.

30. As noted previously, the category of "sect" was particularly and damagingly reified under outside occupation of Iraq, as well as turned differently into a political entity. While sect was previously a somewhat fluid and flexible category under Saddam Hussein's secular pan-Arab Iraqi Sunnism, it became far more rigid. This transformation occurred in particular during a series of American politico-military moves upon occupying Iraq: during U.S. de-Ba'athification policies, the Iraqi Army was dissolved, and the U.S. military increasingly mobilized Shia and Kurdish militias to quash insurgency in the country. Resistance that was previously anti-imperialist (and nonsectarian) took on a sectarian cast. According to Nir Rosen, "The American sectarian approach has created the civil war [in Iraq]. We saw Iraqis as Sunnis, Shias, and Kurds. We designed a governing council based on a sectarian quota system and ignored Iraqis (not exiled politicians but real Iraqis) who warned us against it. We decided that the Sunnis were the bad guys and the Shia were the good guys. These problems were not timeless. In many ways, they are new, and we are responsible for them" (Rosen 2014). These categories were salient for my interlocutors, who remarked on how things had changed from a time when "once we were all just Iraqi."

31. Kali Rubaii describes how sectarianism "becomes a social fact through the repetition and performance of a merger between sect and place in the global war on terror" (2019, 126). She describes an evolution into a state of what she calls "tripartheid: a three-way apartheid system in which people can expect different sets of rights guaranteed in different territories, based on imposed ethno-sectarian categories." Most crucially, although imposed from the outside, tripartheid becomes "internal to being . . . through the forced repetition of imposed categories" (128).

CHAPTER 3. THE THEATERS OF WAR

1. See the discussion of Timothy Mitchell's "world as exhibition" (1989) later in the chapter.

2. Tom Boellstorf disputes "the false opposition between the digital and the real" as "misrepresenti[ng] the relationship between the physical and those phenomena referred to by terms like 'digital,' 'online,' or 'virtual'" (2016, 1). He shows how online experiences are *real* (e.g., if you lose money gambling online, you've lost money in the actual world), while elements of the concrete physical world are more "unreal, as in forms of play and fantasy." Boellstorff critiques scholars like Christine Hines and Sherry Turkle for presupposing such an analytic binary.

3. Role-players received variable contracts. Some would have regular contracts a few days a month or even a few weeks a month. Others had contracts every few months. I heard of some contracts where role-players were paid $350 a day, including food and access to a shower at a hotel, and other instances where they were paid $220, or in some instances $250, without these extra benefits. Contracts in the earlier years of the simulations were often higher. Also, some families were able to get contracts for most days a month, making the job quite lucrative; whereas others only had intermittent contracts. These discrepancies produced tensions between the role-players, as some proposed certain families were favored and others out of that circle.

4. A simulation may be understood here as "a bounded action that bears performative reference to another action, which is or stands to become more legitimate or weighty in another time or context" (Magelssen 2014, 4).

5. Scott Magelssen contends that the theater of the simulations generates "a ludic, ritualized space of play, where identities are performatively forged and stabilized through reenactment, or pre-enactment, on the limen of the actual Iraq— also a complex of performative choices" (2009, 63). As Magelssen suggests, identities are, on a certain level, performed and stabilized within the discrete spaces of military performances themselves. In his book *Simming: Participatory Performance and the Making of Meaning*, Magelssen focuses in particular on spaces of simulation with valences of pedagogy or of social change—becoming spaces, alternately, of verifying and cementing norms; or of imagining alternatives.

6. According to that official, "Staff in the resettlement agencies in the U.S. are bound to hire someone who *won't hurt them*, over someone who is qualified." He noted further that: "The fracturing of the Iraqi community in the U.S. has been detrimental to their resettlement. Even the Bosnians, Croats, and Serbs helped each other out more. The Iraqis are afraid of turning each other in." Indeed, he noted that their fears were not entirely unwarranted; once a month, his resettlement organization received what he called a "poison pen letter, interrupting a person's resettlement file, that would say: the person you are processing [for emigration] is part of the Mahdi Army, etc."

7. There may be a deeper history with the idiom in Iraq as well, given the heft of the term *eating* in a country that endured a decade of sanctions and thus times of hunger. Thank you to Kali Rubaii for this insight.

8. Idioms of destructive consumption were also used to describe damaging forms of talk about others. Gossip is also sometimes described as "*eating* someone's good traits" (*ta'akul ḥasanāthu*), a riff off of the Qu'ranic verse that states that gossiping is equivalent to eating the flesh of one's dead brother; that is, the most intimate desecration: "O you who believe! avoid most of suspicion, for surely suspicion in some cases is a sin, and do not spy nor let some of you backbite others. Does one of you like to eat the flesh of his dead brother? But you abhor it; and be careful of (your duty to) Allah, surely Allah is Oft-returning (to mercy), Merciful" (49:12, Shakir translation of the Qu'ran).

9. For example, those with key roles like the mayor or imam receive more detailed scripts, whereas villagers or mourning mothers receive more generalized guidelines.

10. According to Homi Bhabha, there is a double temporality in the production of national affiliations: the people are both the "historical objects of a nationalist pedagogy" and the "subjects of a process of signification." Through this process, " [t]he scraps, patches, and rags of daily life must be repeatedly turned into the signs of a national culture, while the very act of the narrative performance interpellates a growing circle of national subjects" (1990, 297). The space of simulations generates an unusually heightened nationalist pedagogy.

11. Timothy Mitchell, in his discussion of the panoramic spectacle of the World's Fair, describes the impulse in nineteenth-century Europe around turning the world into an exhibition. For Mitchell, a consequence of orientalism's ordering of the world's objects is that: "The world is grasped, inevitably, in terms of a distinction between the object—the *thing itself* as the European says—and its meaning, with no sense of the historical peculiarity of this effect we call the *thing itself* (1989, 232). Mitchell cites Bourdieu's analogies between the anthropologist, tourist, and the orientalist: "In terms of this distinction the scholar can grasp the world as an exhibition, as a representation, 'in the sense of idealist philosophy, but also as used in painting or the theatre,' and people's lives appear as no more than 'stage parts . . . or the implementing of plans'" (232).

12. Gayatri Spivak points to the possibility of self-conscious, useful, and strategic enactments of affiliation, describing "a strategic use of positivist essentialism in a scrupulously visible political interest" (1988, 205). For Spivak, amidst Western imperialism, minority or subaltern groups of some kind might deploy temporary and instrumental solidarities or identities as a means of political action, even amidst and despite disparities between the groups.

13. Wide scholarship on the veil has described its complex dimensions, from religious and moral obligation to political performance and resistance to imperialism to social norms, to means of juggling the contemporary world and tradition at once, and what Franz Fanon (1952) calls the "historic dynamism of the veil." In his documentary, "On Boys, Girls, and the Veil," Yousry Nasrallah pushes against the notion that the veil is linked exclusively to religion, describing its social dimensions (modesty, fashion, peer pressure, etc.). Fatima Mernissi conversely locates Moroccan judgment of female sexuality and justification for the veil within the religious order. Fadwa El Guindi historicizes the return to the veil (by young urban college students) in the 1970s as part of a new social and political movement and the "liberation from imposed, imported identities, consumerist behaviors" (1999, 184). Many thinkers show the contextual and potentially performative possibilities of the veil: Abu Lughod describes the act of veiling as one "undertaken by women to express their virtue in encounters with particular categories of men" (1986, 159). Hoffman-Ladd proposes that the veil enables women to be "modern and Islamic at the same time" (1987, 40), while MacLeod suggests a similar tug and further proposes that the veil facilitates the movement of working-class women in the public sphere. Mahmood contends that the wearing of the veil is constitutive, part of cultivating religious interiority, and almost like a performance for the self. Other recent writings on the veil have also shown its multivalent possibilities (Almila and Inglis 2017; Shirazi 2018).

14. Goffman asserts that games are "in principle, devoid of important repercussions upon the solidity and continuity of collective and institutional life" (1961, 17). However, his tract often explores the very limits of this principle. Goffman is equally interested in the kinds of anomie and disruption when "the minute social system of face-to-face interaction breaks down" (1959, 12). The military and its contractors construct and imagine the war game's box with very particular parameters and rules of relevance. According to Goffman, the game's rules demarcate what is considered "real" within that context.

15. Goffman asserts that all games have "inhibitory rules, that tell participants what they must not attend to" and "facilitating rules, that tell them what they must recognize." Of the rules of irrelevance, Goffman explains: "It appears that whether checkers are played with bottle-tops on a piece of squared linoleum, with gold figurines on inlaid marble, or with uniformed men standing on colored flagstones in a specially arranged court square, the pairs of players can start

with the 'same' positions, employ the same sequence of strategic moves and coun-
termoves, and generate the same contours of excitement" (1961, 19–20).

16. Jean Baudrillard posits a distinct historical evolution in the human con-
strual of the relation between objects and images, the sign and the signified,
reality and symbols (relying on a European arc of history to develop his model).
In the premodern period, the image functioned as a place-holder for the object, a
transparent representation of the original. After the Industrial Revolution and
the incipient development of capitalism, distinctions between the image and the
object began to disintegrate in the wake of mass production. The proliferation of
images—copies—posed threats to the integrity and primacy of the original
object. Baudrillard's focus, however, is in the stage that he suggests we now
inhabit: the postmodern age of simulation, whereby we are in the realm of the
image alone. The image not only precedes the original, but entirely dissolves the
distinction between the two, and the ability to adjudicate between them. Thus,
"the map precedes the territory" and the sign acts as the "death sentence of every
reference" (1983, 11): that is, signs of the real are substituted for the real itself.
In this sense, for example, Americans encountering the Jungle Cruise ride at
Disneyland, with its parade of threats along its banks—natives with spears,
lions, etc.—issue a death sentence to Africa. However, the majority of those visi-
tors to Disneyland were probably not in Africa prior to experiencing the ride.

17. Virilio contends that the "history of battle is primarily the history of radically changing fields of perception," proposing that the soldier's experience has
progressively become more disembodied, altering relationship between image
and object (1989, 10). Virilio focuses on how the body becomes increasing prosthetic and the eye increasingly mediated as "image was starting to gain sway over
the object, time over space," suggesting that target areas had increasingly come
to resemble cinema location (1). Virilio tracks this phenomenon over time, asking us, for example, to imagine the soldiers of World War I as "spectators of a
pyrotechnic fairy-play . . . a continuous performance, all day all night" (88) and
suggesting that during World War II, Hitler-as-director introduced synthetic
sound effects into his control room. For Virilio, ever-increasing commingling
between image and object, the potential and the actual, and anticipated and the
experienced, creates an effect such that the "the young army recruit cannot truly
imagine what war is like, for a moment, still thinks he is at a show" (61).

18. Der Derian situates Eisenhower's original military industrial complex
within a broader confluence of forces, including the media and entertainment
networks: the "Military-Industrial-media-entertainment or MIME Net." Zigza-
gging through various "unreal" environments implicated in this network, he visits (among others) combat simulations at Fort Irwin with sets developed in Hollywood; U.S. NATO forces enacting Bosnia peace games in Germany; and
America's largest Modeling, Simulation, and Training Conference in Orlando,

Florida, hosted by the U.S. military and attended by weapons industry CEOs and Disney architects alike.

19. In *Performing Remains*, Rebecca Schneider describes how Civil War reen-actors may experience something like "syncopated time . . . where *then* and *now* punctuate each other" (2011, 2). Although predeployment simulations are future rather than past oriented, the paradigm is relevant at least for the role-players, who already have intimate knowledge of "there" and "then." Schneider is inter-ested in the "warp and draw of one time in another"—such that even though reenactment is "not the thing itself (the past), it is somehow also not not the thing" (6–8). Elsewhere, Mark Auslander posits that "touching the past" may be catalyzed through material vectors such as replicas and props, as he describes of a man acting out being sold on an auction block: "It was him, and it wasn't him" (2013, 162).

20. In an article on a lynching reenactment, Auslander (2010) proposes that this commingling of times creates something like a heterotopia, a possibility for historical redress. We come to apprehend, in disturbing intimacy, those who are dead, and just about to die, those who murder and those who suffer, those who come back to life, and those forever consigned to the outer darkness. We are given a chance to move among the unspoken nightmare of our common history.

CHAPTER 4. LEFT AND RIGHT LIMITS

1. In their essay "The Culture Industry: The Enlightenment as Mass Decep-tion" in *The Dialectic of Enlightenment,* Adorno and Horkheimer contend that new forms of mass media and entertainment had subsumed "culture" into a mar-ket logic. They assert: "the film star with whom one is supposed to fall in love is, from the start, a copy of himself"—that is, from the outset, slotted into Holly-wood archetypes (2002, 112). Adorno and Horkheimer tell us of Kant's anticipa-tion of Hollywood: "images are precensored during production by the same standard of understanding which will later determine reception by viewers" (65). Indeed, whereas in the Kantian model, subjects were required to actively classify objects within a priori categories, the Hollywood culture industry "already con-tains the process of identifying, cataloguing and classifying" (104). In this, no contribution is required: "the spectator must need no thoughts of his own" (109), and everything is serial, already a copy of itself. Mass-produced culture of this ilk renders its audience docile and passive, convinced that their needs can only be met by market capitalism, and unable to critique the system itself.

2. See the props and services offered by Opfor Solutions, http://opforsolutions.com/atmospherics.

3. Indeed, the cultural industry mandates that "every phenomenon is by now so thoroughly imprinted by the schema that nothing can occur that does not bear

in advance the trace of the jargon" (2002, 101). Adorno and Horkheimer propose that "for the consumer there is nothing left to classify" (98).

4. Scott Magelssen makes particular use of the term *fidelity* in his study of Fort Irwin: "Fort Irwin strove to reproduce the martial Iraqi landscape with 'fidelity,' but necessarily compressed time and space to expose soldiers to as many crises as it could in their brief training period" (2014, 162). He also uses the military term *texture* to describe authenticating details, for example: "While we waited, he told us about some issues of 'texture' he had been working on. He recently found out that the tea set they thought they'd been using in their bilateral talks ('bilats') with Iraqis was actually a Turkish coffee set, and they had to find a replacement" (166). Magelssen and Davis confirm my own observations, that authenticity is privileged whenever possible and when it does not interfere with operational goals in trainings.

5. In her discussion of Cold War civil defense simulations, Tracy C. Davis notes: "realism is not only selectively deployed, it is selectively desired" (2007, 74). She explains that realism and authenticity aren't always the key goals of simulations, as they may conflict with other pedagogical goals in any given exercise. This conflict likewise manifested in American military trainings. Davis describes how exercise planners sought moments of resemblance as "anchors to legitimize 'real world' applicability" (72), but ultimately had many other factors to juggle. She elaborates: "What civil defense professionals call realistic . . . is not subject to scrutiny as a coherent aesthetic but is subject to analysis as elements that bear special meaning within the exercises, within historically limited time frames, and for specific cultural, institutional, or organizational participant communities" (75). For example, time can be sped up or slowed down for the participants, in order to encompass a more pedagogically valuable arc.

6. Emily Sogn has focused on the Army's institutionalization of "resilience" (2014) as an attribute understood to be fostered through trainings. Beatrice Jauregui analyzes the military logics that seek to produce "responsible, resilient, even invincible subjects" focusing on "world fitness" as a means to make all humanity safe, led by the U.S. Army (2015, 452). See also work on "the dark side" of soldiers' comprehensive programs (Eidelson, Pilisuk, and Soldz 2011) and Alison Howell's notion of resilience as "a schema for the operation of politics" (2015).

7. The acquisition of this way of knowing might be described as somewhere between gaining tacit knowledge alongside a range of more explicit knowledge. As Michael Polanyi explained in his book *The Tacit Dimension:* "We know more than we can tell" ([1966] 2009, 4). Indeed, as a U.S. Army Research Institute Report for the Behavioral and Social Sciences noted: "tacit knowledge is defined as knowledge grounded in experience, intimately related to action" (Hedlund 1999, i). In addition, soldiers are schooled quite explicitly and actively in information about the region of deployment.

8. See also MacLeish (2012; 2015) on technologies like armor interfacing with and habituating to the warzone, especially regarding how this kind of attunement is refined through discipline and experience in ways that are both anticipatory and reactive, as well as Bickford (2019) on how the soldier is formed as "kill-proof," creating a new kind of "survivability."

9. See "Situational Training Exercises," for more general information, www .globalsecurity.org/military/ops/stx.htm. There are examples of STX lane exercises used by the conventional army described elsewhere, such as www.army.mil /article/86952/search_and_attack_stx_training_lane_scenario_replicate_real_ life_combat_situations_. Scott Magelssen described these exercises and the range of cultural scenarios and dilemmas they present in detail in *Simming: Participatory Performance and the Making of Meaning* (2014). For documentation of the Soldiers Urban Reaction Facility (SURF), constructed 2004–2007, see Robert W. Jones Jr., "Camp MacKall: A History of Training," *Veritas* 3, no. 4 (2007): https:// arsof-history.org/articles/pdf/v3n4_camp_mackall.pdf. The situational training exercises were used by Civil Affairs and Psychological Operations officers and NCOs, where "the scenarios allow the cadre to evaluate individual and collective performance in a controlled, but realistic operational environment" (27). Jones describes the use of four scenarios, what I have described as the quadrant.

10. This process of determining social actors was integral to the more basic simulations and also crucial in the initial phases of the more complex simulations.

11. In *A Thousand Plateaus* (1987), Deleuze and Guattari imagine a closed equation that demarcates identity as essence, such that every term, x, is explicitly marked by not being an Other (that is, y). This equation excludes the possibility for contingency and fluctuation in personhood (1987).

12. The term FTX is used broadly as "Field Training Exercise," described as a "high-cost, high-overhead exercise where the entire battalion and its supporting combat support and combat service support units deploys to field locations to conduct tactical operations under simulated combat conditions" in "Field Training Exercises," www.globalsecurity.org/military/ops/ftx.htm. For other descriptions of the FTX, as following the SURF exercises for CA and PSYOP, see Robert W. Jones Jr., "Camp MacKall: A History of Training," *Veritas* 3, no. 4 (2007): https://arsof-history.org/articles/pdf/v3n4_camp_mackall.pdf, which is also a "series of adaptive learning scenarios."

13. Brochure procured at Fort Irwin, National Training Center, California.

14. In the foreword to *A Thousand Plateaus* (1987, xiii), Brian Massumi proposes a more contingent equation, which incorporates successive terms, each which alter the texture and outcome—producing largely unknowable effects.

15. An effects based approach to operations (EBAO) orients operations around outcomes. See for example this U.S. Air Force doctrine explanation: "An indirect effect is a second-order, third-order, or higher-order effect created through an intermediate effect or causal linkage following a tactical action—usually a

delayed and/or displaced consequence associated with the action that caused the direct effect(s)." www.doctrine.af.mil/Portals/61/documents/Annex_3-60 /3-60-D05-Target-EB.

16. Magelssen writes of these complex simulations: "These acts fit into an immense puzzle that the Americans, if they did everything 'right,' could piece together" (2014, 160). That is, there is no radical contingency outside this scheme; if they apply themselves and don't make errors, the American military can make any world knowable.

17. In *Orientalism*, Edward Said's watershed critique of Western representations of the East, he contended that the perils of orientalism come from "disregarding, essentializing, denuding the humanity of another culture, people, or geographic region. But Orientalism has taken a further step than that: it views the Orient as something whose existence is not only displayed but has remained fixed in time and place for the West" (1977, 108). For Said, in this analytic, the East resided in synchronic essences, origins, and pure identities, what he called "synchronic essentialism," which posited that "the whole Orient can be seen panoptically" (1977, 240). Later postcolonial thinkers invoked the importance of taking hybridity, plurality, discontinuity, and globalization into consideration (i.e., Bhabha 1991, Appadurai 1996). Meanwhile, Foucauldian critiques excavated the links between knowledge and power, while Michel-Rolph Trouillot developed the concept of the savage slot (1991), where the Other was construed as an inverse that helped the West construe its own face.

18. In particular, the projects of nineteenth-century anthropologists, hired by colonial governments to collect knowledge about their colonies, were rebuked, as seeking to categorize those cultures in their synchronic and "pure" forms, purged of historical and colonial accretions. In *Culture and Truth*, Renato Rosaldo critiques the "Lone Ethnographer" for representing "the colonized as members of a harmonious, internally homogenous, unchanging culture" proposing that the "'timeless traditional culture' served as a self-congratulatory reference point against which civilization could measure its own progressive evolution" (1993, 31). In the same period, James Clifford rebuked an analytic approach that saw the world as "populated by endangered authenticities" (1988, 5). Critical anthropologists and social and literary theorists refuted the notion that other societies existed as bounded totalities. Thinkers of identity and nationalism noted the impulse to recover a community's authenticity through a shared past. Partha Chatterjee (1993) describes how "inner" spiritual domains were construed as inviolable in nation-making projects. In an article on authenticity's ambiguities, Julie Skurski notes that "authenticity" has been linked "to ideas of undisputed origins, original creation, and sustained tradition, rather than to notions of imitation, appropriation, and syncretism" (1994, 611).

19. In this sense, the position of the role-players echoes a colonial logic of "cultural translation," wherein the actual knowledges of the Other are not

necessarily pertinent. As I have noted, cultural translation enacts fantasies of mediation, while being enmeshed in power relations (Asad 1984) and potentially dilutes and commensurates radical worlds (Povinelli 2001). In this case, the particularity of the Middle Eastern culture is fed into the military's preexisting schemas and archetypes.

20. In his fieldwork at Fort Irwin, Scott Magelssen describes "a healthy camaraderie between Iraqis and American NTC staff there, as well as between members of different Iraqi sects and ethnicities, which was not to be found everywhere in the real Iraq. Together, the American and Iraqi actors produced and coproduced narratives about Iraq in the shared space of the Sandbox, each group with its own complex agenda, each responding to given situations from a dynamic habitus, and each within the mediated flux of controlled free play. Admittedly, my impressions were carefully mediated by Fort Irwin itself. I don't entertain illusions that I saw anything the army *didn't* want me to see" (2014, 175). He then goes on to describe how the Iraqi role-players he interacted with candidly condemned the U.S. handling of the war in Iraq (however, leading them to express support of the trainings framed as beneficial, and thereby essentially following the military line) and concluded that "the performances that took place at Fort Irwin cannot easily be framed and dismissed as either a manipulative objectification and exoticization of Iraqis on display or a patriotic bit of military pageantry, serving purely ideological ends" (175).

21. Indeed, in his *The Principles of Scientific Management*, Frederick Taylor (1911) described how scientific principles could be applied to industry, coordinating the system to maximize benefits for all (including increasing wages for laborers). Key here is that Taylor emphasizes how a worker's body, mind, and inner life had to be disciplined in order to maximize their productivity.

22. I observed a summer program of training cadets who were still in school and not en route in any immediate sense to deploy. Those trainings also contracted role-players, but the environment was more relaxed than the official predeployment trainings. The incident with "Lubna" occurred at such a training. However, the subsequent example with the cultural advisor, "Ghaith" occurred at a normal predeployment training.

23. In "Can the Subaltern Speak," Gayatri Spivak writes, "White men are saving brown women from brown men" to describe the British colonial attempt to eradicate suttee. More broadly, it might be viewed as one of the key (and gendered) modes of relation between the colonizer and the colonized ([1995] 2006, 33). See also Leila Ahmed's writings (1992) on how "the woman question" is construed in colonialism, wherein oppression of Muslim women is caused by timeless patriarchal Muslim "culture," positioned in opposition to modern (and white) Western European culture's respect for women, and Lila Abu Lughod's (2002) "Do Muslim Women Really Need Saving? Anthropological Reflections on Cultural Relativism and Its Others."

24. In Magelssen's account, Iraqi role-players sought to avoid "bad guy" roles, as they were pro-American and wanted to distance themselves from such positions (2014, 13). I saw a much wider range of experiences and training outcomes among the Iraqis I interviewed.

25. Thanks to Ken MacLeish for this helpful point.

CHAPTER 5. AFFECTIVE MANEUVERS

1. According to Judith Butler, precarity "designates that politically induced condition in which certain populations suffer from failing social and economic networks of support and become differentially exposed to injury, violence, and death" (2009, 25). As Butler explains, "there are 'subjects' who are not quite recognizable as subjects, and there are "lives" that are not quite—or, indeed, are never—recognized as lives" (4). This condition "underscores our radical substitutability and anonymity" (14). Indeed, for Butler, war brings this bifurcation into the relief: "We might think of war as dividing populations into those who are grievable and those who are not. An ungrievable life is one that cannot be mourned because it has never lived, that is, it has never counted as a life at all" (38). As I have discussed elsewhere, Iraqi intermediaries have lived in the locus of the accused from all sides, and thus within exposure and particular precarity. Irrespective of these Iraqis' commitments and motivations for their wartime choices, the critical gaze of both other Iraqis and American soldiers frequently projected potential duplicity and inauthenticity upon them; in consequence, they were often the first to be killed by militias, often mistrusted and not sufficiently protected by the soldiers they worked for.

2. When power relations make certain identities, lives, or worlds unthinkable and thereby ungrievable, Butler describes this condition as nothing short of a psychic wound producing melancholia, "a refusal of grief and an incorporation of loss, a miming of the death it cannot mourn" (1997, 142).

3. This role-player was particularly interested in the thirteenth-century Gnostic text, "The Book of the Sun" (Kitab Shams al-Ma'araf), which is composed of magical squares, meant to be portals to the jinn. The text had been banned under Saddam Hussein, and together we read sections of it, which I had procured from my university library. Billel explained that the jinn can seize and enter the human being and send him into the black room, and that conversely, in order to make the jinn do your bidding, you had to strengthen yourself. We looked in the index of the book for a reference to the "black room," so he could explain the moment in the text to me. We did not find the black room in the index, but instead, we found "the black spirits" and "the black analyses." Billel compared this sensation of warding off the "black room" to a feeling he had experienced during the war games.

4. Freud writes of a sensation of involuntarily circling back: "As for instance when one is lost in a forest in high altitudes, caught, we will suppose, by the

mountain mist, and when every endeavor to find the marked or familiar path ends again and again in a return to one and the same spot, recognizable by some particular landmark" ([1919] 1963, 43). The representations of death and injury, and of the Middle East, create just such an impression of uneasy familiarity, commingled with strangeness, as something intensely familiar is uprooted and resituated.

5. Frederic Jameson describes the "pastiche" as an imitation of something idiosyncratic yet "amputated of the satiric impulse. Pastiche is . . . blank parody. A statue with blind eyeballs" (1991, 17).

6. As I described in the introduction, I did not witness any instances where soldiers were trained in any Arabic dialect.

7. See Hillel Cohen, *Good Arabs: The Israeli Security Agencies and the Israeli Arabs, 1948–1967* (2010), about Palestinian collaborators. In an inquiry that is equal parts about compliance and resistance, Cohen describes the lives and motivations and structural circumstances of the subset of the Palestinian population who worked with the Israelis. Deemed "their" Arabs, "the Israeli authorities filled leadership positions in the Arab community with 'moderates'—the official term for Arabs who refrained from taking nationalist positions and who accorded legitimacy not only to Israel's existence but also to its actions, such as the imposition of military rule over the Arab population and the expropriation of Arab lands" (3).

8. Butler explains: "If subversion is possible, it will be subversion from within the terms of the law, through the possibilities that emerge when the law turns against itself and spawns unexpected permutations of itself" (1999, 119). The law in this instance might be understood as the archetypal face imagined by military scriptwriters, a face required to repeat itself so many times that it becomes mechanical, turning the person into a cultural tool.

9. In her monograph on laughter, Anca Parvulescu asks what "work or (unwork) does laughter do?" and "what kind of subjects are we when we laugh?" (2010, 3). For Georges Bataille, it is the "dramatization" of existence, which catalyzes the kind of inner experience that prompts laughter: "If we hadn't known how to dramatize, we wouldn't know how to laugh, but in us laughter is always ready to stream forth into a renewed fusion" ([1954] 1988, 11). The fusion Bataille describes here intimates a proximity to dying, something akin to the subject-position of the role-players, positioned in between recent wartime vulnerability and feigned death.

10. The laughter Bataille describes is well contextualized by Derrida's discussion of potential economies of the body. Derrida explains what he calls the Hegelian "restricted economy": "Hegel clearly had proclaimed the necessity of the master's retaining of the life that he exposes to risk" ([1967] 1978, 255). According to Derrida, Bataille, in contrast, creates a "general economy," where in risking death, rather than trying to make sense or meaning out of it, he laughs. For

Bataille, laughter is prompted by being on the cusp of the kind of undifferentiation that death precipitates: a fusion with the void. Yet in a crucial turn, Bataille describes the potency of becoming proximate to death, *without dying:* "In question is the death of an *other,* but in such cases the death of the other is always the image of one's own death" (Bataille in Derrida [1967] 1978, 258). For example, in a wake, the observer cannot but temporarily inhabit the corpse, but still escapes that body's fate. So too in Bataille's archetypal example, in the spectacle of sacrifice, to observe is to project the self into the locus of the obliterated body without being obliterated.

11. For Bataille, this living death generates an intensified inner experience, where the subject is for a moment stranded—between his discrete autonomy and his disintegration into the all. This moment of theater, in which the subject enacts death while still alive, entails an output of energy without a satisfying recuperation of sense or meaning, causing the jolt of laughter. Derrida explains how the consequences of the dramatization: "It must simulate, after a fashion, the absolute risk, and it must laugh at this simulacrum" ([1967] 1978, 256). In Bataille's conceit, the person at the edge projects himself into the deceased body of the other. In contrast, the role-player actually embodies that body. In the mime of the simulations, many of the role-players who are former interpreters accused of collaboration and previously earmarked for death are the participant-observers in the spectacles of their own deaths. Living in between the recent risk of death and the continuous feint of death, they offer a heightened case-study of Bataille's concept.

12. Recently, some anthropologists have read affect as a means of "creating new subjects and relations between subjects," in some instances offering the potential for "contemporary political and economic transformations" (Richard and Rudnyckyj 2009, 62). Others have pointed instead to affect's role in driving social reproduction and reinforcing norms. For example, Lauren Berlant argues that affects of belonging precipitate individuals to reproduce normative forms of patriarchy and femininity (2008) and Vincanne Adams describes how post-Katrina forms of sociality transforms affect into a source of profit, reproducing entrenched structures of inequality (2013).

13. Recent segments of John Oliver's show—about America's immigration courts, which forces children to appear alone in court, and a segment about Trump's family separation policy—help to further crystallize how the affect is functioning. Oliver explained that because immigration cases are not considered criminal cases, the American government is not legally compelled to offer lawyers to defendants—even children. Oliver then replies incredulously: "That's just clearly ridiculous. You cannot let a two-year-old be unsupervised in court. You can't even let the two-year-old be unsupervised in a bouncy castle. They're going to come out covered in glitter, holding a broken beer bottle and a dead bird." See www.youtube.com/watch?v=ygVX1z6tDGI and www.youtube.com/watch?v=9fBOGBwJ2QA.

CHAPTER 6. BECOMING HUMAN TECHNOLOGY

1. To clarify, as part of their own formation, the Green Beret trainees are training the guerillas to fight. That is, the American soldier trainees are in effect training how to train. The guerillas are all contracted role-players, and some of them are retired military.

2. During Robin Sage, there are four key parts: first, classes on Unconventional Warfare fundamentals. Second, a mission rehearsal exercise, where students engage with a series of "dilemma lanes," engaging in dilemmas such as "meet a guerilla chief, attend a sector command meeting and conduct a key leader engagement" (Woytowich 2016, 31). Third, in an isolation faculty where they conduct mission planning. Fourth, in a twelve-day field training exercise in Pineland. I attended a portion of part four.

3. When culminating exercises for the Green Berets launched in 1952, they were staged in the Chattahoochee National Forest in Georgia. Previously known as "Gobbler Woods" and "Cherokee Trail," the exercise was renamed Robin Sage after it moved near Robins, North Carolina (and also as a tribute to former Army colonel Jerry Sage, a veteran of World War II and an Office of Strategic Services officer who trained soldiers in Unconventional Warfare). Different iterations of Robin Sage have been written about by Douglas Waller (*The Commandos: The Inside Story of America's Secret Soldiers*, 1994); Tom Clancy (*Shadow Warriors: Inside the Special Forces*, 2002); Fred Pushies (*U.S. Army Special Forces*, 2001); Gerald Schumacher (*To Be a U.S. Army Green Beret*, 2005); Dick Couch (*Chosen Soldier: The Making of a Special Forces Warrior*, 2007); Linda Robinson (*Masters of Chaos: The Secret History of the Special Forces*, 2009); Tony Schwalm (*The Guerilla Factory: The Making of Special Forces Officers, The Green Berets*, 2013), and others.

4. See also Woytowich, "The Relevancy of Robin Sage," *Special Warfare* (July–Dec. 2016) for more examples of early Pineland history and role-players who have been involved since childhood.

5. Eyal Weizman traces such striation through the Israeli Defense Force's (IDF) strategic and tactical appropriation of Deleuze and Guattari's "war-machine," wherein the IDF theorizes and executes the "smoothing of space" by "walking through walls" (2006). Weizman shows how the state seeks to *mimic and appropriate* the nomadic war-machine: "They [the IDF] used none of the city's streets, roads, alleys or courtyards, or any of the external doors, internal stairwells and windows, but moved horizontally through walls and vertically through holes blasted in ceilings and floors. This form of movement, described by the military as 'infestation,' seeks to redefine inside as outside, and domestic interiors as thoroughfares. The IDF's strategy of 'walking through walls' involves a conception of the city as not just the site but also the very medium of warfare—a flexible, almost liquid medium that is forever contingent and in flux" (2006).

6. This emphasis on the siloing of different insurgencies into trainings (i.e., into ethnicity, sect, politics, criminality) rather than bringing together their intersectionality, seemed like a particular idealization and distancing from war on the ground. Thank you to Ken MacLeish for this insight.

7. There is an intricate backstory here, which goes back to 1870 and is detailed in *Atlantica: A Concise History,* authored by Johns Hopkins Applied Physics Research Lab in 2012 in partnership with the 1st Battalion, 1st Special Warfare Training Group (Airborne). While I had access to this document during my research, it is not available in the public domain, so I have not included further details.

8. The role-players at the site I was at were a combination of civilian contractors playing native Pinelanders and military manpower. Iraqis role-played guerillas at other sites during this particular training.

9. I was unable to take notes on the spot during this encounter and thus all quotes are approximations from memory.

10. The killing of a rabbit in military predeployment training perhaps has some degree of precedent. In one account of the Vietnam War (Andrew Hunt's *The Turning: A History of Vietnam Veterans Against the War,* 1999), a veteran described how training activities at Camp Pendleton had prepared soldiers to commit acts of atrocity in Vietnam. He describes how an officer produced a rabbit, allowed the soldiers to develop an attachment to it, then killed it, skinned it, and disemboweled it in front of them. Thank you to Nadia Abu El-Haj for sharing this material.

11. For example, Derek Gregory writes: "The emphasis on cultural difference— the attempt to hold the Other at a distance while claiming to cross the interpretative divide—produces a diagram in which violence has its origins in 'their' space, which the cultural turn endlessly partitions through its obsessive preoccupation with ethno-sectarian division, while the impulse to understand is confined to 'our' space, which is constructed as open, unitary and generous: the locus of a hermeneutic invitation that can never be reciprocated" (2008, 18).

12. In his 2007 critique of the military's Cultural Turn in "Counterpunch," David Price accuses military anthropology of providing "the military with just the sort of support, rather than illumination, that they seek." He castigates the military for "looking for basic courses in local manners so they can get on with the job of conquest" while "ignor[ing] anthropological critiques of power, colonialism, militarization, hegemony, warfare, cultural domination" (2007).

13. E. Valentine Daniel writes: "The desire to find culture, either as a present reality or as a deferred ideal, to find it in any case, as a coherent whole, true and beautiful, is a desire to find a corpse." He explains: "The problem lies at the core of the culture concept itself" (1996, 200–201).

CONCLUSION

1. The Pentagon has conducted recent ethics reviews of the Special Operations community (in 2018, 2019, and 2020) in response to an array of alleged war crimes. These include: an alleged execution of a fifteen-year-old ISIS prisoner in 2017, after which Navy SEAL Edward Gallager posed with the body; the alleged strangling of a Green Beret by two Navy SEALs and two Marines in Mali in 2017; an alleged murder and cover-up of an unarmed Afghan man by the Green Beret Matthew Golsteyn in 2010; and the alleged rape of two children by a Special Forces soldier in 2017. For more details about the incidents that prompted the initial review, see: www.businessinsider.com/dod-is-reviewing-special-operations-forces-after-horrible-incidents-2018-12?r=US&IR=T.

2. For example, Michael Taylor and his son, former Special Forces soldiers, were accused of helping to smuggle former Nissan chief executive Carlos Ghosn out of Osaka, Japan; Ghosn was facing financial crimes in Japan and under strict bail conditions at the time. In the second instance, former Special Forces operators became entangled in a Venezuela coup, an operation the United States did not support. To note, the U.S. denied having a role in this operation, and the military was not otherwise present. See https://edition.cnn.com/2020/05/20/politics/ghosn-smuggling-arrest/index.html and https://sofrep.com/news/the-plot-thickens-as-two-ex-special-forces-operators-are-caught-in-venezuela-coup.

3. The 2020 review was conducted after the court martial of a Navy SEAL platoon chief, Edward Gallagher, on allegations of murdering a captive and other war crimes in Iraq in 2017, which ended in acquittal of all major charges. This review followed congressionally mandated reviews of the Special Operations' culture and accountability in 2018 and its ethics in 2019, which were not properly implemented, according to the report. See www.nytimes.com/2020/01/29/us/eddie-gallagher-us-military-special-forces.html.

EPILOGUE

1. For example: focusing on the "vulnerability of the observer" (Behar 1996); the impact of being a "halfie" anthropologist (Abu Lughod 1991); weaving poetics into epistemologies (Crapanzano 1992); incorporating the art of interlocutors (Biehl 2013); an interest in "blurred genres" (Behar 2007).

2. To name a few: Taussig (1987, 2003); Behar (1993, 1996, 2008); Gottlieb and Graham (1994); Narayan (1989, 2004); Jackson (1986, 1995); Stewart (2007); Stoller (1989, 1997, 1999); Raffles (2010); Kusserow (2013); Rosaldo (2013); Tsing (2015).

3. Kent Maynard proposed that anthropology and poetry had gone through parallel historical trajectories in the twentieth century, as both disciplines became suspicious of totalizing structures (2002). He also reflected on poetry as a epistemological mode, "help[ing] us to know more about society" (2009, 115). Maynard described how anthropologists might attend to poetry as one kind of epistemology—i.e., learning to "position [themselves] better as ethnographers within a scene," "attend[ing] to the whole person," and focusing on affect, which at its best is "an impassioned pursuit of the real" (2009, 121–25). He cautioned that whatever genre we write in, we must attend to questions of the "factual"—i.e., what is knowable and how (125).

4. See Philip Metres, "From Reznikoff to Public Enemy," *Poetry Foundation*, www.poetryfoundation.org/articles/68969/from-reznikoff-to-public-enemy.

5. A wide range of additional contemporary poets have used field or interview methods of some kind, loosely defined. See Susan Rich, Joe Hall, Loretta Collins Klobah, Julia Spicer Kasdorf, Joshua Gottleib-Miller, Tracy Zeman, Yerra Sugarman, Jennifer Jean, Orchid Tierney, John Canaday, and many others.

6. Thank you for the poet Gretchen Marquette's brilliant conversations and insights on how to restructure *Kill Class*.

7. More recently, Kent Maynard has noted the importance of sound in poetry, noting that both poets and anthropologists "remark on the centrality of rhythm and rhyme to social life" (2009, 119), yet without delving into how this might be enacted in ethnographic poetics.

8. Frost used this phrase in a lecture in 1915 at the Browne and Nichols School.

Bibliography

Abu Lughod, Lila. 1986. *Veiled Sentiments: Honor and Poetry in a Bedouin Society.* Berkeley: University of California Press.

———. 1989. "Zones of Theory in the Anthropology of the Arab World." *Annual Review of Anthropology* 18:267–306.

———. 1991. "Writing against Culture." In *Recapturing Anthropology: Working in Present,* edited by Richard Foxed. Santa Fe, NM: School of American Research Press.

———. 2002. "Do Muslim Women Really Need Saving? Anthropological Reflections on Cultural Relativism and Its Others." *American Anthropologist* 104 (3): 783–90.

Aburish, Said. 2002. "How Saddam Hussein Came to Power." In *The Saddam Hussein Reader,* edited by Turi Munthe. New York: Thunder's Mouth Press.

Achebe, Chinua. 1988. "The Truth of Fiction." In *Hopes and Impediments: Selected Essays,* 138–52. New York: Doubleday.

Adams, Vincanne. 2013. *Markets of Sorrow, Labors of Faith: New Orleans in the Wake of Katrina.* Durham, NC: Duke University Press.

Adorno, Theodor. 2005. *Minima Moralia: Reflections on a Damaged Life.* New York: Verso.

Adorno, Theodor, and Max Horkheimer. 2002. *The Dialectic of Enlightenment.* Stanford, CA: Stanford University Press.

Ahmed, Leila. 1992. *Women and Gender in Islam: Historical Roots of a Modern Debate.* New Haven, CT: Yale University Press.

Al-Ali, Nadje Sadig. 2007. *Iraqi Women: Untold Stories from 1948 to the Present*. London: Zed Books.

Albro, Robert, and Bill Ivey, eds. 2014. *Cultural Awareness in the Military: Developments and Implications for Future Humanitarian Cooperation*. London: Palgrave Macmillan UK.

Al-Bukhari, Muhammad. [2009]. *Sahiha al-Bukhari*. Translated by Muhammad Muhsin Khan. Houston: Dar-us-Salam Publications.

Ali, Zahra. 2018. *Women and Gender in Iraq: Between Nation-Building and Fragmentation*. Cambridge: Cambridge University Press.

Al-Khalil, Samir [Makiya Kanan]. 1989. *Republic of Fear: The Politics of Modern Iraq*. Berkeley: University of California Press.

Allen, Robertson. 2017. *America's Digital Army: Games at Work and War*. Lincoln: University of Nebraska Press.

Almila, Anna-Mari, and David Inglis. 2017. *The Routledge Handbook to Veils and Veiling*. New York: Routledge.

Al-Mohammad, Hayder. 2010. "Towards an Ethics of Being-With: Intertwinements of Life in Post-invasion Basra." *Ethnos: Journal of Anthropology* 75 (4): 425–46.

———. 2011. "'You Have Car Insurance, We Have Tribes': Negotiating Everyday Life in Basra and the Re-emergence of Tribalism." *Anthropology of the Middle East* 6 (1): 18–34.

———. 2012. "A Kidnapping in Basra: The Struggles and Precariousness of Life in Postinvasion Iraq." *Cultural Anthropology* 27, no. 4 (November): 597–614.

———. 2019. "What Is the 'Preparation' in the Preparing for Death?" *Current Anthropology* 60 (6): 796–812. https://doi.org/10.1086/706756.

Al-Mohammad, Hayder, and Daniela Peluso. 2012. "Ethics and the 'Rough Ground' of the Everyday: Overlappings of Life in Postinvasion Iraq." *HAU: Journal of Ethnographic Theory* 2 (2): 42–59.

American Anthropological Association Executive Board. 2007. "Statement on the Human Terrain System Project." www.americananthro.org/ConnectWith AAA/Content.aspx?ItemNumber=1626.

Appadurai, Arjun. 1996. *Modernity at Large: Cultural Dimensions of Globalization*. Public Worlds Book 1. Minneapolis: University of Minnesota Press.

Apter, Emily. 2006. *The Translation Zone: A New Comparative Literature*. Princeton, NJ: Princeton University Press.

———. 2014. "Translation at the Checkpoint." *Journal of Postcolonial Writing* 50 (1): 56–74.

Arendt, Hannah. 1976. *The Origins of Totalitarianism*. New York: Harcourt.

Asad, Talal. 1973. *Anthropology and the Colonial Encounter*. New York: Ithaca Press.

———. 1983. "Anthropological Conceptions of Religion: Reflections on Geertz." *Man* 18 (2): 237–59.

———. 1986. "The Concept of Cultural Translation." In *Writing Culture*, edited by James Clifford and George Marcus. Berkeley: University of California Press.

———. 1992. "Conscripts of Western Civilization." In *Dialectical Anthropology: Essays in Honor of Stanley Diamond*, edited by C. W. Gailey. Tallahassee: University Press of Florida.

Ashcroft, Bill, Gareth Griffiths, and Helen Tiffin. [1989] 2003. *The Empire Writes Back: Theory and Practice in Post-Colonial Literatures.* 2nd ed. London: Routledge.

Auslander, Mark. 2010. "'Holding on to Those Who Can't Be Held': Reenacting a Lynching at Moore's Ford, Georgia." *Southern Spaces.* http://southernspaces .org/2010/holding-those-who-cant-be-held-reenacting-lynching-moores-ford-georgia.

———. 2013. "Touching the Past: Materializing Time in Traumatic 'Living History' Reenactments." *Signs and Society* 1 (1): 161–83.

Baiocchi, Dave. 2013. "Measuring Army Deployments to Iraq and Afghanistan." Rand Corporation. www.rand.org/pubs/research_reports/RR145.html.

Bakhtin, Michel. 1984. *Rabelais and His World.* Bloomington: Indiana University Press.

Barakat, Sultan. 2005. "Post Saddam Iraq: Deconstructing a Regime, Reconstructing a Nation." *Third World Quarterly* 26 (4–5): 571–91.

Barkawi, Tarak, and Ketih Stanski. 2013. *Orientalism and War.* London: Hurst.

Barth, Frederik. [1969] 1998. *Ethnic Groups and Boundaries: The Social Organization of Cultural Difference.* Long Grove, IL: Waveland Press.

Bassnett, Susan, and André Lefevere, eds. 1990. *Translation, History and Culture.* London: Pinter.

Bassnett, Susan, and Harish Trivedi, eds. 1999. *Postcolonial Translation: Theory and Practice.* London: Routledge Press.

Bataille, Georges. [1954] 1988. *Inner Experiences.* Translated by Leslie Anne Boldt. Albany: SUNY Press.

Batatu, Hanna. 1978. *The Old Social Classes and the Revolutionary Movements of Iraq: A Study of Iraq's Old Landed and Commercial Classes and Its Communists, Ba'thists, and Free Officers.* Princeton, NJ: Princeton University Press.

Bateson, Gregory. 1961. "Fun in Games." In *Encounters: Two Studies in the Sociology of Interaction*, edited by Erving Goffman. New York: Bobs-Merrill.

———. 1972. "A Theory of Play and Fantasy." In *Steps to an Ecology of the Mind.* Chicago: University of Chicago Press.

Baudrillard, Jean. [1981] 1995. Translated by Sheila Glaser. *Simulacra and Simulation.* Ann Arbor, MI: University of Michigan Press.

———. 1983. *Simulations.* New York: Semiotext(e).

Beech, Christopher. 2013. "Behind the Green Beret: Building Rapport." *Special Forces News (SOFREP)*, September 10. https://sofrep.com/26411/green-beret-ussoc-army-sf.

Behar, Ruth. 1993. *Translated Woman: Crossing the Border with Esperanza's Story*. Boston: Beacon Press.

———. 1996. *The Vulnerable Observer: Anthropology That Breaks Your Heart*. Boston: Beacon Press.

———. 2007. "Ethnography in a Time of Blurred Genres." *Anthropology and Humanism* 32 (2): 145–55.

———. 2008. "The Portable Island: Cubans at Home in the World." New York: Palgrave.

Benedict, Ruth. 1946. *The Chrysanthemum and the Sword: Patterns of Japanese Culture*. New York: Houghton Mifflin.

Benjamin, Medea. 2012. *Drone Warfare: Killing by Remote Control*. New York: Verso.

Benjamin, Walter. [1968] 2007. *Illuminations: Essays and Reflections*. Edited by Hannah Arendt. Translated by Harry Zohn. New York: Schocken Books.

Berg, Ulla, and Ana Y. Ramos-Zayas. 2015. "Racializing Affect: A Theoretical Proposition." *Current Anthropology* 56 (5): 654–77.

Bergen, Peter, and Daniel Rothenberg, eds. 2014. *Drone Wars: Transforming Conflict, Law, and Policy*. New York: Cambridge University Press.

Bergson, Henri. [1900] 1914. *Laughter: An Essay on the Meaning of the Comic*. Translated by Cloudesley Brereton and Fred Rothwell. New York: Macmillan.

Berlant, Lauren. 1997. *The Queen of America Goes to Washington City: Essays on Sex and Citizenship*. Durham, NC: Duke University Press.

———. 2008. *The Female Complaint: The Unfinished Business of Sentimentality in American Culture*. Durham, NC: Duke University Press.

Bernal, Victoria. 2013. "Please Forget Democracy and Justice: Eritrean Politics and the Powers of Humor." *American Ethnologist* 40, no. 2: 300–309.

Bhabha, Homi. 1983. "The Other Question . . . Homi Bhabha Reconsiders the Stereotype and Colonial Discourse." *Screen* 24.

———. 1990. "DissemiNation: Time, Narrative, and the Margins of the Modern Nation." In *Nation and Narration*, edited by Homi Bhabha, 291–322. London: Routledge.

———. 1991. "'Race,' Time, and the Revision of Modernity." *Oxford Literary Review* 13 (1): 193–219.

———. 1994. *The Location of Culture*. London: Routledge.

Bickford, Andrew. 2019. "'Kill-Proofing' the Soldier: Environmental Threats, Anticipation, and the US Military Biomedical Armor Program." *Current Anthropology* 60 (S19): S39–S48.

Biehl, Joao. 2013. *Vita: Life in a Zone of Social Abandonment*. Berkeley: University of California Press.

Birtle, Andrew. 2006. *US Army Counterinsurgency and Contingency Operations Doctrine 1942–1976*. Amazon Digital Service.

Blackburn, Stuart. 2006. *Print, Folklore, and Nationalism in Colonial Southern India*. New Delhi: Pauls Press.

Boellstorf, Tom. 2008. *Coming of Age in Second Life: An Anthropologist Explores the Virtually Human*. Princeton: Princeton University Press.

———. 2016. "For Whom the Ontology Turns: Theorizing the Digital Real." *Current Anthropology* 57 (4): 387–98.

Bourdieu, Pierre. [1984] 2010. *Distinction*. Abingdon: Routledge and Kegan Paul.

Boyer, Dominic, and Alexei Yurchak. 2010. "American Stiob: Or, What Late-Socialist Aesthetics of Parody Reveal about Contemporary Political Culture in the West." *Cultural Anthropology* 25, no. 2: 179–221.

Braidotti, Rosi. 2013. *The Posthuman*. Cambridge: Polity Press.

Brandel, Andrew. 2018. "A Poet in the Field: The Companionship of Literature and Anthropology." *Anthropology of This Century* 21. http://aotcpress.com/articles/a-poet-in-the-field.

Bubandt, Nils, and Rane Willerslev. 2015. "The Dark Side of Empathy: Mimesis, Deception and the Magic of Alterity." *Comparative Studies in Society and History* 57 (1): 5–34.

Butler, Judith. [1990] 1999. *Gender Trouble: Feminism and the Subversion of Identity*. New York: Routledge.

———. 1997. *The Psychic Life of Power*. Stanford, CA: Stanford University Press.

———. 2004. *Precarious Life: The Powers of Mourning and Violence*. London: Verso.

———. 2009. *Frames of War*. London: Verso.

Byrd, Jodi. 2011. *The Transit of Empire: Indigenous Critiques of Colonialism*. Minneapolis: University of Minnesota Press.

Cahnmann-Taylor, Melisa. 2016. *Imperfect Tense*. San Pedro, CA: Whitepoint Press.

Campbell, Madeline Otis. 2016. *Interpreters of Occupation: Gender and the Politics of Belonging in an Iraqi Refugee Network*. Syracuse, NY: Syracuse University Press.

Canguilhem, Georges. 1992. "Machine and Organism." In *Incorporations*, edited by Jonathan Crary and Sanford Kwinter, 44–69. Translated by Mark Cohen and Randall Cherry. New York: Zone Books.

Chamayou, Grégoire. 2015. *A Theory of the Drone*. Translated by Janet Lloyd. New York: The New Press.

Chapman, Anne. 2010. *The Origins and Development of the National Training Center, 1976–1984*. Washington, DC: Office of the Command Historian, United States Army and Doctrine Command. https://history.army.mil/html /books/069/69-3/CMH_Pub_69-3.pdf.

Chatterjee, Partha. 1993. *The Nation and Its Fragments: Colonial and Postcolonial Histories*. Princeton, NJ: Princeton University Press.

Choi, Don Mee. 2020. *DMZ Colony*. Seattle: Wave Books.

Chomsky, Noam. 2008. "Humanitarian Imperialism: The New Doctrine of Imperial Right." *Monthly Review*, Sept. 1. https://chomsky.info/200809___.

Chow, Rey. 2010. "The Age of the World Target: Atomic Bombs, Alterity, Area Studies." In *The Rey Chow Reader*. New York: Columbia University Press.

Clifford, James. 1988. *The Predicament of Culture: Twentieth-Century Ethnography, Literature and Art*. Cambridge, MA: Harvard University Press.

Clifford, James, and George Marcus. 1986. *Writing Culture: The Poetics and Politics of Ethnography*. Berkeley: University of California Press.

Coburn, Noah. 2018. *Under Contract: The Invisible Workers of America's Global War*. Stanford, CA: Stanford University Press.

Cockburn, Patrick. 2007. *The Occupation: War and Resistance in Iraq*. New York: Verso.

Cohen, Hillel. 2010. *Good Arabs: The Israeli Security Agencies and the Israeli Arabs, 1948–1967*. Berkeley: University of California Press.

Colla, Elliott. 2015. "Dragomen and Checkpoints." *The Translator* 21 (2): 132–53.

Connable, Ben. 2009. "All Our Eggs in a Broken Basket." *Military Review*, April 6. www.army.mil/article/19281/all_our_eggs_in_a_broken_ basket_how_the_human_terrain_system_is_undermining_sustainable_ military_cu.

Conrad, Joseph. [1899] 1998. *Heart of Darkness*. Edited by Nicholas Tredell. New York: Columbia University Press.

Couch, Dick. 2008. *Chosen Soldier: The Making of a Special Forces Warrior*. New York: Three Rivers Press.

Crapanzano, Vincent. 1992. *Hermes' Dilemma and Hamlet's Desire*. Cambridge, MA: Harvard University Press.

Cudworth, Erika. 2015. "Killing Animals: Sociology, Species Relation and Institutionalized Violence." *Sociological Review* 63 (1): 1–18.

Dabashi, Hamid. 2011. *Brown Skin, White Masks*. London: Pluto Press.

Daniel, Errol Valentine. 1996. *Charred Lullabies*. Princeton, NJ: Princeton University Press.

Davis, Rochelle. 2010. "Culture as a Weapon System." *Middle East Report* 255:9–13.

Davis, Tracy C. 2007. *Stages of Emergency: Cold War Nuclear Civil Defense*. Durham, NC: Duke University Press.

Deleuze, Gilles, and Félix Guattari. 1987. *A Thousand Plateaus: Capitalism and Schizophrenia*. Minneapolis: University of Minnesota Press.

De Man, Paul. 1986. *The Resistance to Theory*. Minneapolis: University of Minnesota Press.

Dennison, Jean. 2012. *Colonial Entanglement: Constituting a Twenty-First-Century Osage Nation*. Chapel Hill: University of North Carolina Press.

Der Derian, James. 2009. *Virtuous War: Mapping the Military Industrial Entertainment Complex*. New York: Routledge.

Derrida, Jacques. [1967] 1978. *Writing and Difference*. Translated by Alan Bass. New York: Routledge.

———. 1985a. "Des Tours de Babel." In *Difference in Translation*, edited by Joseph F. Graham, 165–207. Ithaca, NY: Cornell University Press.

———. 1985b. *The Ear of the Other*. Edited by Christie McDonald. Translated by Avita Ronell and Peggy Kamuf. New York: Schocken.

de Waal, Frans. 2007. "The 'Russian Doll' Model of Empathy and Imitation." In *On Being Moved: From Mirror Neurons to Empathy*, edited by Stein Bråten, 49–69. Amsterdam: John Benjamins.

Dewachi, Omar. 2017. *Ungovernable Life: Mandatory Medicine and Statecraft in Iraq*. Stanford, CA: Stanford University Press.

DiPrizio, Robert. 2002. *Armed Humanitarians: US Interventions from Northern Iraq to Kosovo*. Baltimore, MD: Johns Hopkins University Press.

Dodge, Toby. 2003. *Inventing Iraq*. New York: Columbia University Press.

Durkheim, Emile. [1893] 1984. *The Division of Labor in Society*. New York: The Free Press.

Edwards, Paul. 1996. *The Closed World: Computers and the Politics of Discourse in Cold War America*. Cambridge, MA: MIT Press.

Eickelman, Dale. [1981] 2001. *The Middle East and Central Asia: An Anthropological Approach*. New York: Prentice Hall.

Eidelson, Roy, Marc Pilisuk, and Stephen Soldz. 2011. "The Dark Side of Comprehensive Soldier Fitness." *American Psychologist* 66 (7): 643–44.

El Guindi, Fadwa. 1999. *Veil: Modesty, Privacy, and Resistance*. New York: Berg Publishers.

Elyachar, Julia. 2005. *Markets of Dispossession: NGOs, Economic Development, and the State in Cairo*. Durham, NC: Duke University Press.

Faizullah, Tarfia. 2014. *Seam*. Minneapolis, MN: Graywolf Press.

Fanon, Frantz. [1952] 2008. *Black Skin, White Masks*. Translated by Richard Philcox. New York: Grove.

Fassin, Didier, and Mariella Pandolfi. 2010. "Introduction: Military and Humanitarian Government in the Age of Intervention." In *Contemporary States of Emergency: The Politics of Military and Humanitarian Interventions*, edited by Didier Fassin and Mariella Pandolfi, 9–25. New York: Zone Books.

Feldman, Ilana. 2015. *Police Encounters: Security and Surveillance in Gaza under Egyptian Rule.* Stanford, CA: Stanford University Press.

Finley, Erin. 2011. *Fields of Combat: Understanding PTSD among Veterans of Iraq and Afghanistan.* Ithaca, NY: Cornell University Press.

Fischer, Hannah. 2013. "U.S. Military Casualty Statistics: Operation New Dawn, Operation Iraqi Freedom, and Operation Enduring Freedom." Congressional Research Service Report for Congress, February 5. https://journalistsresource.org/wp-content/uploads/2013/02/RS22452.pdf.

Flynn, Michael, James Sisco, and David Ellis. 2012. "Left of Bang: The Value of Sociocultural Analysis in Today's Environment." *Prism: A Journal of the Center for Complex Operations* 3 (4): 13–21.

Forte, Maximilian, ed. 2010. *The New Imperialism.* Vol. 1, *Militarism, Humanism, and Occupation.* Montreal: Alert Press.

———. 2011. "The Human Terrain System and Anthropology." *Public Anthropology* 113 (1): 149–53.

Fosher, Kerry. 2014. Foreword in *Cross Cultural Competence for a Twenty-First Century Military,* edited by Allison and Robert Greene-Sands. Plymouth, UK: Lexington Books.

Foster, Robert. 2001. "Unvarnished Truths: Maslyn Williams and Australian Government Film in Papua and New Guinea." In *In Colonial New Guinea: Anthropological Perspectives,* edited by Naomi M. McPherson, 64–82. Pittsburgh: University of Pittsburgh Press.

Foucault, Michel, 1977. *Discipline and Punish: The Birth of the Prison.* New York: Pantheon Books.

———. [1997] 2003. *"Society Must Be Defended": Lectures at the Collège de France.* Edited by Mauro Bertani and Alessandro Fontana. Translated by David Macey. New York: Picador.

———. 2001. *Power (The Essential Works of Foucault, 1954–1984, Vol. 3).* Edited by James Faubian. Translated by Robert Hurley. New York: The New Press.

———. 2007. *"Security, Territory, Population": Lectures at the Collège de France, 1977–1978.* Edited by Michel Senellart, Francois Ewald, Alessandro Fontana, and Arnold Davidson. Translated by Graham Burchell. New York: Palgrave Macmillan.

Frazer, James George. [1980] 2009. *The Golden Bough: A Study in Magic and Religion.* Oxford: Oxford University Press.

Freud, Sigmund. [1919] 1963. "The Uncanny." In *Studies in Parapsychology.* Translated by Alix Strachey. Edited by Philip Rieff. New York: Collier Books, 19–62.

———. 1960. *Jokes and Their Relation to the Unconscious.* New York: Norton.

Gant, Jim. 2009. "A Strategy for Success in Afghanistan: One Tribe at a Time." Los Angeles: Nine Sisters Imports. www.globalsecurity.org/military/library /report/2009/2009_one_tribe_at_a_time.pdf.

Geertz, Clifford. 1973. *The Interpretation of Cultures*. New York: Basic Books.

Gentile, Gian. 2013. *Wrong Turn: America's Deadly Embrace of Counterinsurgency*. New York: New Press.

Gillen, Francis, and Baldwin Spencer. 1899. *Native Tribes in Central Australia*. New York: Macmillan.

Giordano, Cristiana. 2014. *Migrants in Translation: Caring and Logics of Difference in Contemporary Italy*. Oakland: University of California Press.

Goffman, Erving. 1959. *Presentation of Self in Everyday Life*. New York: Doubleday.

———. 1961. "Fun in Games." In *Encounters: Two Studies in the Sociology of Interaction*. Harmondsworth: Penguin.

Goldstein, Donna M. 2003. *Laughter Out of Place: Race, Class, Violence, and Sexuality in a Rio Shantytown*. Berkeley: University of California Press.

Goldstein, Joseph. 2015. "U.S. Soldiers Told to Ignore Sexual Abuse of Boys by Afghan Allies." *New York Times*. September 20. www.nytimes.com/2015 /09/21/world/asia/us-soldiers-told-to-ignore-afghan-allies-abuse-of-boys .html?_r=0.

Golomski, Casey. 2019. "Breathing Room: Poetic Form as Resistance to Convention in the Ethnography of Suffering." *Somatosphere*, March 8. http:// somatosphere.net/2019/breathing-room-poetic-form-as-resistance-to-convention-in-the-ethnography-of-suffering.html.

González, Roberto. 2009. *American Counterinsurgency: Human Science and the Human Terrain*. Chicago: Prickly Paradigm Press.

———. 2013. "Cybernetic Crystal Ball: 'Forecasting' Insurgency in Iraq and Afghanistan." In *Virtual War and Magical Death: Technologies and Imaginaries for Terror and Killing*, edited by Neil L. Whitehead and Sverker Finnström, 65–84. Durham, NC: Duke University Press.

———. 2016. *Militarizing Culture: Essays on the Warfare State*. New York: Routledge.

Gottlieb, Alma, and Philip Graham. 1994. *Parallel Worlds: An Anthropologist and a Writer Encounter Africa*. Chicago: University of Chicago Press.

Graham, Steven. 2007. "War and the City." *New Left Review* 44 (March/April): 121–33.

Gregory, Derek. 2004. *The Colonial Present: Afghanistan * Palestine * Iraq*. Maldon, MA: Blackwell Publishing.

———. 2008. "The Rush to the Intimate: Counterinsurgency and the Cultural Turn in Late Modern Warfare." *Radical Philosophy* 150:8–23.

Grossman, Dave. 2009. *On Killing: The Psychological Cost of Learning to Kill in Wartime*. New York: Back Bay Books.

Guattari, Félix. 1995. *Chaosmosis: An Ethico-Aesthetic Paradigm*. Bloomington: Indiana University Press.

Gusterson, Hugh. 2007. "Anthropology and Militarism." *Annual Review of Anthropology* 36:155–75.

———. 2016. *Drone: Remote Control Warfare*. Cambridge, MA: MIT Press.

Haddad, Fanar. 2011. *Sectarianism in Iraq: Antagonistic Visions of Unity*. Oxford: Oxford University Press.

Halpern, Jodi. 2001. *From Detached Concern to Empathy: Humanizing Medical Practice*. Oxford: Oxford University Press.

Hameed, Basim. 2008. "Islamic Jihad in Iraq: Suicide or Martyrdom." In *Suicide Bombers: The Psychological, Religious, and Other Imperatives*, edited by Mary Sharpe, 135–43. Amsterdam: IOS Press.

Hamilton Cushing, Frank. [1879] 2011. *Participant Observation: A Guide for Fieldworkers*. 2nd ed. Edited by Kathleen DeWalt and Billie DeWalt. New York: Altamira Press.

Haraway, Donna. 1991. "A Cyborg Manifesto: Science, Technology, and Socialist-Feminism in the Late Twentieth Century." In *Simians, Cyborgs and Women: The Reinvention of Nature*, 149–81. New York: Routledge.

Hardt, Michael, and Antonio Negri. 2000. *Empire*. Cambridge, MA: Harvard University Press.

Harman, Graham. 2002. *Tool Being: Heidegger and the Metaphysics of Objects*. Chicago: Open Court Publishing.

Harrison, Faye V., ed. 1991. *Decolonizing Anthropology: Moving Further Toward an Anthropology for Liberation*. Arlington, VA: American Anthropological Association.

Harvey, David. 2003. *The New Imperialism*. Oxford: Oxford University Press.

Hastings, Michael. 2012. "Bowe Berghdal: America's Last Prisoner of War." *Rolling Stone*, June 7. www.rollingstone.com/politics/politics-news/bowe-bergdahl-americas-last-prisoner-of-war-189891.

Hayles, Katherine N. 1999. *How We Became Posthuman: Virtual Bodies in Cybernetics, Literature, and Informatics*. Chicago: University of Chicago Press.

Hedlund, Jennifer. 1999. *Tacit Knowledge for Military Leaders: Platoon Leader's Questionnaire*. U.S. Army Research Institute for Behavioral and Social Sciences Fort Leavenworth Research Unit. https://apps.dtic.mil/dtic/tr/fulltext/u2/a362346.pdf.

Heidegger, Martin. [1953] 1996. *Being and Time*. Translated by Joan Stambaugh. Albany: SUNY Press.

———. [1954] 1977. *The Question Concerning Technology*. Translated by William Lovitt. New York: Harper Perennial.

Hoffman, David. 2000. *Empathy and Moral Development: Implications for Caring and Justice*. Cambridge: Cambridge University Press.

Hoffman-Ladd, Valerie. 1987. "Polemics on the Modesty and Segregation of Women in Contemporary Egypt." *International Journal of Middle East Studies* 19 (1): 23–50.

Holland, Douglass, and C. Jason Throop. 2011. *An Anthropology of Empathy*. New York: Berghahn Books.

Holmes-Eber, Paula, and Barak A. Salmoni. 2008. *Operational Culture for the Warfighter*. Quantico, VA: Marine Corps University Press.

Holmes-Eber, Paula, Patrice M. Scanlon, and Andrea L. Hamlen. 2009. *Applications in Operational Culture: Perspectives from the Field*. Quantico, VA: Marine Corps University Press.

Howe, David. 2012. *Empathy: What It Is and Why It Matters*. New York: Palgrave.

Howell, Alison. 2015. "Resilience as Enhancement: Governmentality and Political Economy beyond 'Responsibilisation.'" *Politics* 35 (1): 67–71.

Huhndorf, Shari Michelle. 2001. *Going Native: Indians in the American Cultural Imagination*. Ithaca, NY: Cornell University Press.

Hunt, Andrew. 1999. *The Turning: A History of Vietnam Veterans against the War*. New York: New York University Press.

Inghilleri, Moira. 2010. "You Don't Make War without Knowing Why: The Decision to Interpret in Iraq." *Translator* 16 (2): 175–96.

Ivy, Marilyn. 2008. "Benedict's Shame." *Cabinet* 31 (Fall), www.cabinetmagazine .org/issues/31/ivy.php?source=post_page.

Jackson, Gary. 2021. *Origin Story*. Albuquerque: University of New Mexico Press.

Jackson, Michael. 1986. *Barawa and the Way Birds Fly in the Sky*. Washington, DC: Smithsonian Institution Press.

———. 1995. *At Home in the World*. Durham, NC: Duke University Press.

Jameson, Frederic. 1991. *Postmodernism, or, the Cultural Logic of Late Capitalism*. Durham, NC: Duke University Press.

Jauregui, Beatrice. 2015. "World Fitness: US Army Family Humanism and the Positive Science of Persistent War." *Public Culture* 27 (3): 449–85.

Jawad, Saad. 2006. "Iraqi Intellectuals and the Occupations." Interview by Laith Al-Saud. *Counterpunch*, January 3. www.counterpunch.org/2006 /01/03/iraqi-intellectuals-and-the-occupation.

Jeffrey, James E. 2015. "Why Counterinsurgency Doesn't Work." *Foreign Affairs* (March/April). www.foreignaffairs.com/articles/united-states/2015-02-16 /why-counterinsurgency-doesnt-work.

Kanaaneh, Rhoda Ann. 2008. *Surrounded: Palestinian Soldiers in the Israeli Military*. Stanford, CA: Stanford University Press.

Kaplan, Fred. 2014. *The Insurgents: David Petraeus and the Plot to Change the American Way of War*. New York: Simon and Schuster.

Kelly, John, Beatrice Jauregui, Sean T. Mitchell, and Jeremy Walton, eds. 2010. *Anthropology and Global Counterinsurgency*. Chicago: University of Chicago Press.

Kelly, Tobias, and Sharika Thiranagama, eds. 2010. *Traitors: Suspicion, Intimacy, and the Ethics of State-Building*. Philadelphia: University of Pennsylvania Press.

Khalili, Laleh. 2012. *Time in the Shadows: Confinement in Counterinsurgency*. Stanford, CA: Stanford University Press.

Khanna, Ranjana. 2003. *Dark Continents: Psychoanalysis and Colonialism*. Durham, NC: Duke University Press.

Kilcullen, David. 2006. "Twenty-Eight Articles: Fundamentals of Company-Level Counterinsurgency." *Iosphere: Joint Operations Information* Center (Summer): 29–35. www.pegc.us/archive/Journals/iosphere_summer06_kilcullen.pdf.

———. 2007. "Religion and Insurgency." *Small Wars Journal Blog*. May 12. https://smallwarsjournal.com/blog/religion-and-insurgency.

Kilshaw, Susie. 2009. *Impotent Warriors: Gulf War Syndrome, Vulnerability and Masculinity*. New York: Berghahn.

Kipp, Jacob, Lester Grau, Karl Prinslow, and Captain Don Smith. 2006. "The Human Terrain System: A CORDS for the 21st Century." *Military Review* (September–October): 8–15.

Klein, Naomi. 2007. *The Shock Doctrine: The Rise of Disaster Capitalism*. New York: Picador.

Kohn, Richard H. 2009. "The Danger of Militarization in an Endless 'War' on Terrorism." *Journal of Military History* 73 (1): 177–208.

Krznaric, Roman. 2015. *Empathy: Why It Matters, and How to Get It*. New York: TarcherPerigee.

Kusserow, Adrie. 2013. *Refuge*. Rochester, NY: BOA Editions.

Lakoff, Andrew. 2007. "Preparing for the Next Emergency." *Public Culture* 19 (2): 247–71.

Lawrence, T. E. [1922] 2003. *The Seven Pillars of Wisdom*. Whitefish, MT: Kessinger Publishing.

Levi, Primo. 1988. *The Drowned and the Saved*. New York: Vintage.

Levin, Dana. 2011. "Where It Breaks: Drama, Silence, Speed, Accrual." In *A Broken Thing: Poets on the Line*, edited by Emily Rosko and Anton Vander-Zee. Iowa City: University of Iowa Press.

Longenbach, James. 2004. *The Resistance to Poetry*. Chicago: University of Chicago Press.

———. 2018. *How Poems Get Made*. New York: W. W. Norton.

Lukács, Georg. 1971. *History and Class Consciousness: Studies in Marxist Dialectics*. Translated by Rodney Livingstone. Cambridge, MA: MIT Press.

Lutz, Catherine. 2001. *Homefront: A Military City and the American Twentieth Century.* Boston: Beacon Press.

———. 2006. "Empire Is in the Details." *American Ethnologist* 33 (4): 593–611.

———. 2009. "Anthropology in an Era of Permanent War." *Anthropologica* 51 (2): 367–79.

MacLeish, Kenneth. 2012. "Armor and Anesthesia: Exposure, Feeling, and the Soldier's Body." *Medical Anthropology Quarterly* 26 (1): 49–68.

———. 2015. *Making War at Fort Hood: Life and Uncertainty in a Military Community.* Princeton, NJ: Princeton University Press.

MacLeod, Arlene Elowe. 1992. "Hegemonic Relations and Gender Resistance: The New Veiling as Accommodating Protest in Cairo." *Signs* 17 (3): 533–57.

Magelssen, Scott. 2009. "Rehearsing the 'Warrior Ethos': 'Theatre Immersion' and the Simulation of Theatres of War." *Drama Review* 53 (1): 47–72.

———. 2014. *Simming: Participatory Performance and the Making of Meaning.* Ann Arbor: University of Michigan Press.

Mahmood, Saba. 2004. *Politics of Piety: The Islamic Revival and the Feminist Subject.* Princeton, NJ: Princeton University Press.

Makdisi, Ussama. 2000. *The Culture of Sectarianism: Community, History, and Violence in Nineteenth-Century Ottoman Lebanon.* Berkeley: University of California Press.

Malkki, Liisa H. 1995. *Purity and Exile: Violence, Memory, and National Cosmology among Hutu Refugees in Tanzania.* Chicago: University of Chicago Press.

Mamdani, Mahmood. 2005. *Good Muslim, Bad Muslim: America, the Cold War, and the Roots of Terror.* Danvers, MA: Harmony Books.

Marcus, George. 2000. *Para-Sites: A Casebook against Cynical Reason.* Chicago: University of Chicago Press.

Martinez, Luis, and Amy Bingham. 2011. "U.S. Veterans: By the Numbers." *ABC News,* November 11. http://abcnews.go.com/Politics/us-veterans-numbers/story?id=14928136#.

Marx, Karl. [1867] 2007. *Capital: A Critique of Political Economy.* Vol. 1. New York: Cosimo Classics.

Masco, Joseph. 2006. *The Nuclear Borderlands: The Manhattan Project in Post–Cold War New Mexico.* Princeton, NJ: Princeton University Press.

———. 2014. *Theater of Operations: National Security Affect from the Cold War to the War on Terror.* Durham, NC: Duke University Press.

Massad, Joseph. 2007. *Desiring Arabs.* Chicago: University of Chicago Press.

Massumi, Brian. 1996. "The Autonomy of Affect." In *Deleuze: A Critical Reader,* edited by Paul Patton, 217–39. Oxford: Blackwell.

Mauss, Marcel. 1973. "Techniques of the Body." *Economy and Society* 2 (1): 70–88.

Maynard, Kent. 2002. "An 'Imagination of Order': The Suspicion of Structure in Anthropology and Poetry." *Antioch Review* 60 (2): 220–43.

———. 2009. "Rhyme and Reasons: The Epistemology of Ethnographic Poetry." *Etnofoor* 21 (2): 115–29.

Maynard, Kent, and Melisa Cahnmann-Taylor. 2010. "Anthropology at the Edge of Words: Where Poetry and Ethnography Meet." *Anthropology and Humanism* 35 (1): 2–19.

Mazzarella, William. 2009. "Affect: What Is It Good For?" In *Enchantments of Modernity: Empire, Nation, Globalization,* edited by Saurabh Dube, 292–308. New York: Routledge.

———. 2017. *The Mana of Mass Society.* Chicago: University of Chicago Press.

McCartt-Jackson, Sarah. 2018. *Spotlight.* Portland, OR: Airlie Press.

McCoy, Alfred. 2009. *Policing America's Empire: The United States, the Philippines, and the Rise of the Surveillance State.* Madison: University of Wisconsin Press.

McFate, Montgomery. 2005. "The Military Utility of Understanding Adversary Culture." *Joint Force Quarterly* 38 (3): 42–48.

McFate, Montgomery, and Andrea Jackson. 2005. "An Organizational Solution for DOD's Cultural Knowledge Needs." *Military Review* (July–August): 18–21.

McGranahan, Carole. 2010. *Arrested Histories: Tibet, the CIA, and Memories of a Forgotten War.* Durham, NC: Duke University Press.

McGranahan, Carole, and John Collins, eds. 2018. *Ethnographies of U.S. Empire.* Durham, NC: Duke University Press.

Meredith, William. 2002. "Poem about Morning." http://writersalmanac.publicradio.org/index.php?date=2002/05/29.

Mernissi, Fatima. 1975. *Beyond the Veil: Male-Female Dynamics in Modern Muslim Society.* Bloomington: Indiana University Press.

Merton, Robert. 1972. "Insiders and Outsiders: A Chapter in the Sociology of Knowledge." *American Journal of Sociology* 78 (1): 9–47.

Messinger, Seth D. 2010. "Getting Past the Accident: Explosive Devices, Limb Loss, and Refashioning a Life in a Military Medical Center." *Medical Anthropology Quarterly* 24 (3): 281–303.

Metres, Philip. 2007. "From Reznikoff to Public Enemy: The Poet as Journalist, Historian, Agitator." Poetry Foundation, November 5. www.poetryfoundation.org/articles/68969/from-reznikoff-to-public-enemy.

Mikdashi, Maya. 2014. "Sex and Sectarianism: The Legal Architecture of Lebanese Citizenship." *Comparative Studies of South Asia, Africa, and the Middle East* 34 (2): 279–93.

Mitchell, Timothy. 1989. "The World as Exhibition." *Comparative Studies in Society and History* 31 (2): 217–36.

Miyoshi, Masao, and Harry Harootunian. 2002. *Learning Places: The Afterlives of Area Studies.* Durham, NC: Duke University Press.

Mojaddedi, Fatima. 2019. "The Closing Heart, Mouth, Word." *Public Culture* 31 (3): 497–520.

Moore, Robin. [1965] 1999. *The Green Berets*. New York: Moore Hill Publishing.

Nader, Laura. 1972. "Up the Anthropologist: Perspectives Gained from 'Studying Up.'" In *Reinventing Anthropology*, edited by D. Hyms, 284–311. New York: Random House.

Nagl, John. 2002. *Learning How to Eat Soup with a Knife: Counterinsurgency Lessons from Malaya and Vietnam*. Chicago: University of Chicago Press.

Nakash, Yitzak. 1995. *The Shi'is of Iraq*. Princeton, NJ: Princeton University Press.

Napoleoni, Loretta. 2005. *Terror Incorporated: Tracing the Dollars Behind the Terror Networks*. New York: Seven Stories Books.

Narayan, Kirin. 1989. *Storytellers, Saints, and Scoundrels: Folk Narrative in Hindu Religious Teaching*. Philadelphia: University of Pennsylvania Press.

———. 2004. *Love, Stars, and All That*. Berkeley: University of California Press.

Nasrallah, Yousry, dir. 1995. *On Boys, Girls, and the Veil*. Cairo: MISR International Films.

Network of Concerned Anthropologists. 2009. *The Counter-Counterinsurgency Manual*. New York: Prickly Paradigm Press.

Ondiak, Natalie, and Brian Katulis. 2009. "Operation Safe Haven Iraq 2009: An Action Plan for Airlifting Endangered Iraqis Linked to the United States." Center for American Progress.

Packer, George. 2006. "Knowing the Enemy: Can Social Scientists Redefine the 'War on Terror'?" *New Yorker*, December 18.

———. 2007. "Betrayed." *New Yorker*, March 26. www.newyorker.com/magazine /2007/03/26/betrayed-2.

Pandian, Anand, ed. 2019. *A Possible Anthropology: Methods for Uneasy Times*. Durham, NC: Duke University Press.

Pandian, Anand, and Stuart McLean, eds. 2017. *Crumpled Paper Boat: Experiments in Ethnographic Writing*. Durham, NC: Duke University Press.

Parvulescu, Anca. 2010. *Laughter: Notes on a Passion*. Cambridge, MA: MIT Press.

Pels, Peter, and Oscar Salamink. 1999. *Colonial Subjects: Essays on the Practical History of Anthropology*. Ann Arbor: University of Michigan Press.

Pickup, Sharon. 2010. "Military Training: Army and Marine Corps Face Challenges to Address Projected Future Requirements." United States Government Accountability Office, Report to Congressional Committees.

Pinsky, Robert. 2014. *The Sounds of Poetry: A Brief Guide*. New York: Farrar, Straus, Giroux.

Polanyi, Michael. [1966] 2009. *The Tacit Dimension*. Chicago: University of Chicago Press.

Porter, Patrick. 2007. "Good Anthropology, Bad History: The Cultural Turn in Studying War." *Parameters* 37 (2): 45–58.

Povinelli. Elizabeth. 2001. "Radical Worlds: The Anthropology of Incommensurability and Inconceivability." *Annual Review of Anthropology* 30:319–34.

———. 2011. *Economies of Abandonment: Social Belonging and Endurance in Late Liberalism*. Durham, NC: Duke University Press.

Price, David. 2004. *Threatening Anthropology: McCarthyism and the FBI's Persecution of Activist Anthropologists*. Durham, NC: Duke University Press.

———. 2007. "Pilfered Scholarship Devastates General Petraeus's Counterinsurgency Manual." *Counterpunch*. www.counterpunch.org/2007/10/30 /pilfered-scholarship-devastates-general-petraeuss-counterinsurgency-manual-core-chapter-a-morass-of-borrowed-quotes-university-of-chicago-press-badly-compromised-counterinsurgency.

———. 2008. *Archaeological Intelligence: The Use and Neglect of American Anthropology in the Second World War*. Durham, NC: Duke University Press.

———. 2016. *Dual Use Anthropology: Cold War Anthropologists and the CIA*. Durham, NC: Duke University Press.

Rafael, Vicente. 1992. *Contracting Colonialism: Translation and Christian Conversion in Tagalog Society under Early Christian Rule*. Durham, NC: Duke University Press.

Raffles, Hugh. 2010. *Insectopedia*. New York: Vintage Books.

Reno, Joshua O. 2020. *Military Waste: The Unexpected Consequences of Permanent War Readiness*. Oakland: University of California Press.

Reznikoff, Charles. 1975. *Holocaust*. Los Angeles: Black Sparrow Press.

Rhodes, Jeffrey P. 1989. "All Together at Fort Irwin." *Air Force Magazine* 72 (12): 38–47.

Richard, Analiese, and Daromir Rudnyckyj. 2009. "Economies of Affect." *Journal of the Royal Anthropological Institute* 15 (1): 57–77.

Richardson, Rachel. 2011. *Copperhead*. Pittsburgh: Carnegie Mellon Press.

Rosaldo, Renato. 1993. *Culture and Truth*. Boston: Beacon Press.

———. 2013. *The Day of Shelly's Death: The Poetry and Ethnography of Grief*. Durham, NC: Duke University Press.

Rosen, Nir. 2014. "Iraq's Problems Are Not Timeless. The U.S. Is Responsible." *Boston Review*, June 17. https://bostonreview.net/blog/nir-rosen-juan-cole-john-dower-iraq-war-sunni-shia-conflict.

Royle, Nicholas. 2003. *The Uncanny*. Manchester: Manchester University Press.

Rubaii, Kali. 2019. "Tripartheid: How Sectarianism Became Internal to Being in Anbar, Iraq." *PoLar* 42 (1): 125–41.

Rukeyser, Muriel. 1938. "Book of the Dead." In *Route 1*. New York: Covici Friede.

Rumsfeld, Donald. 2002. "Transforming the Military." *Foreign Affairs* 81 (3): 20–32.

Rutherford, Danilyn. 2009. "Sympathy, State Building, and the Experience of Empire." *Cultural Anthropology* 24 (1): 1–32.

Said, Edward. 1978. *Orientalism*. London: Penguin.

Saleh, Zainab. 2021. *Return to Ruin: Iraqi Narratives of Exile and Nostalgia*. Stanford, CA: Stanford University Press.

Salibi, Kamal. 1979. "Middle Eastern Parallels: Syria-Iraq-Arabia in Ottoman Times." *Middle Eastern Studies* 15 (1): 70–81.

Sandoval, Chela. 2000. *Methodology of the Oppressed*. Minneapolis: Minnesota University Press.

Sartre, Jean-Paul. [1945] 2008. "What Is a Collaborator?" In *Situations III: The Aftermath of War*. London: Seagull.

Scahill, Jeremy. 2007. *Blackwater: The Rise of the World's Most Powerful Mercenary Army*. New York: Nation Books.

Scales, Robert H. 2006. "Clausewitz and World War IV." *Armed Forces Journal* (July). http://armedforcesjournal.com/clausewitz-and-world-war-iv.

Scheper-Hughes, Nancy. 1993. *Death without Weeping*. Berkeley: University of California Press.

Schneider, Rebecca. 2011. *Performing Remains: Art and War in Times of Theatrical Reenactment*. New York: Routledge.

Schwalm, Tony. 2012. *The Guerilla Factory: The Making of Special Forces Officers, the Green Berets*. New York: Free Press.

Scott, David. 2004. *Conscripts of Modernity: The Tragedy of Colonial Enlightenment*. Durham, NC: Duke University Press.

Sheridan, Mary Beth. 2008. "U.S. Troops in Baghdad Take a Softer Approach." *Washington Post*, November 20. www.washingtonpost.com/wp-dyn/content/article/2008/11/19/AR2008111904440.html?noredirect=on.

Shirazi, Faegheh. 2018. *The Veil Unveiled: The Hijab in Muslim Culture*. Gainesville: University Press of Florida.

Simons, Anna J. 1997. *The Company They Keep: Life Inside the U.S. Army Special Forces*. New York: Avon Books.

Simpson, Audra. 2014. *Mohawk Interruptus: Political Life across the Borders of Settler States*. Durham, NC: Duke University Press.

Singer, P. W. 2003. *Corporate Warriors: The Rise of the Privatized Military Industry*. Ithaca, NY: Cornell University Press.

———. 2009. *Wired for War: The Robotics Revolution and Conflict in the 21st Century*. New York: Penguin.

Sinha, Nitin. 2012. *Communication and Colonialism in Eastern India: Bihar, 1760s–1880s*. London: Anthem Press.

Skultans, Vieda. 2008. *Empathy and Healing: Essays in Medical and Narrative Anthropology*. New York: Berghahn.

Skurski, Julie. 1994. "The Ambiguities of Authenticity in Latin America: Doña Bárbara and the Construction of the National Identity." *Poetics Today* 15 (4): 605–42.

Sluglett, Peter. 2015. "Reflections on Recent Studies in Iraq." In "Roundtable: Perspectives on Researching Iraq Today." *Arab Studies Journal* 23 (1): 259–62.

Sluglett, Peter, and Marion Farouk-Sluglett. 2001. *Iraq Since 1958: From Revolution to Dictatorship*. 3rd edition. New York: Bloomsbury Academic.

Smith, Adam. [1759] 2002. *The Theory of Moral Sentiments*. Cambridge: Cambridge University Press.

Sogn, Emily. 2014. "Throw a Survey at It: Questioning Soldier Resilience." *Anthropology Now*, April 25. http://anthronow.com/print/throw-a-survey-at-it-questioning-soldier-resilience-in-the-us-army.

Soussi, Alasdair. 2010. "Lawrence of Arabia, Guiding US Army in Iraq and Afghanistan." *Christian Science Monitor*. www.csmonitor.com/World/2010/0619/Lawrence-of-Arabia-guiding-US-Army-in-Iraq-and-Afghanistan.

Spivak, Gayatri Chakravorty. 1988. *In Other Worlds: Essays in Cultural Politics*. New York: Routledge.

———. [1995] 2006. "Can the Subaltern Speak?" In *The Post-Colonial Studies Reader*, edited by Bill Ashcroft, Gareth Griffiths, and Helen Tiffin, 28–37. New York: Routledge.

Stewart, Kathleen. 2005. "Cultural Poesis: The Generativity of Emergent Things." In *Handbook of Qualitative Research*, edited by Norman Denzin and Yvonna Lincoln, 1027–42. Thousand Oaks, CA: Sage.

———. 2007. *Ordinary Affects*. Durham, NC: Duke University Press.

Stocking, George. 1991. *Colonial Situations: Essays on the Contextualization of Ethnographic Knowledge*. Madison: University of Wisconsin Press.

Stoler, Ann Laura. 2002. *Carnal Knowledge and Imperial Power: Race and the Intimate in Colonial Rule*. Berkeley: University of California Press.

———. 2009. *Along the Archival Grain*. Princeton, NJ: Princeton University Press.

———. 2016. *Duress: Imperial Durabilities in Our Times*. Durham, NC: Duke University Press.

———. 2018. "Disassemblage: Rethinking U.S. Imperial Formations." In *Ethnographies of U.S. Empire*, edited by Carole McGranahan and John F. Collins. Durham, NC: Duke University Press.

Stoller, Paul. 1989. *The Taste of Ethnographic Things: The Senses in Anthropology*. Philadelphia: University of Pennsylvania Press.

———. 1997. *Sensuous Scholarship*. Philadelphia: University of Pennsylvania Press.

———. 1999. *Jaguar: A Story of Africans in America*. Chicago: University of Chicago Press.

Stone, Nomi. 2008. *Stranger's Notebook*. Chicago: Northwestern University Press / TriQuarterly Books.

———. 2014. "War Game." *Painted Bride Quarterly* 89. http://pbq.drexel.edu /nomi-stone-war-game-america.

———. 2015. "The Door." *Guernica*, May 1. www.guernicamag.com/poetry/the-door.

———. 2017. "Living the Laughscream: Human Technology and Affective Maneuver." *Cultural Anthropology* 32 (1): 149–74.

———. 2018. "Imperial Mimesis: Enacting and Policing Empathy in US Military Training." *American Ethnologist* 45 (4): 533–45.

———. 2019. *Kill Class*. North Adams, MA: Tupelo Press.

Strathern, Marilyn. 1988. *The Gender of the Gift*. Berkeley: University of California Press.

Taussig, Michael. 1987. *Shamanism, Colonialism, and the Wild Man: A Study of Terror and Healing*. Chicago: University of Chicago Press.

———. 1993. *Mimesis and Alterity: A Particular History of the Senses*. New York: Routledge.

———. 2001. "Dying Is Like Art, Like Everything Else." *Critical Inquiry* 28 (1): 305–16.

———. 2003. *Law in a Lawless Land: Diary of Limpieza in Colombia*. Chicago: University of Chicago Press.

———. 2011. *I Swear I Saw This: Drawings in Fieldwork Notebooks, Namely My Own*. Chicago: University of Chicago Press.

Taylor, Frederick. 1911. *The Principles of Scientific Management*. New York: Harper and Brothers Publishers.

Tripp, Charles. 2007. *A History of Iraq*. Cambridge: Cambridge University Press.

Trouillot, Michel-Rolph. 1991. "Anthropology and the Savage Slot: The Poetics and Politics of Otherness." In *Recapturing Anthropology: Working in the Present*, edited by R. G. Fox, 17–45. Santa Fe, NM: School of American Research Press.

Tsing, Anna. 2015. *The Mushroom at the End of the World*. Princeton, NJ: Princeton University Press.

Turkle, Sherry. 1984. *The Second Self: Computers and the Human Spirit*. Cambridge, MA: MIT Press.

Turner, Victor. 1969. *The Ritual Process: Structure and Anti-Structure*. New York: Routledge.

Turse, Nick. 2014. "America's Secret War in 134 Countries." *The Nation*, January 16. www.thenation.com/article/177964/americas-secret-war-134-countries.

Van Devanter, Lynda. [1983] 2001. *Home Before Morning: The Story of an Army Nurse in Vietnam*. Boston: University of Massachusetts Press.

Van Horne, Patrick. 2014. "The Next Level of Tactical Awareness—Getting Left of Bang." *IOA News* (Spring/Summer). www.originalwisdom.com/wp-content/uploads/bsk-pdf-manager/2019/03/Van-Horne_2014_The-Next-Level-of-Tactical-Awareness-getting-left-of-bang.pdf.

Verdery, Katharine. 2018. *My Life as a Spy: Investigations in a Secret Police File.* Durham, NC: Duke University Press.

Vine, David. 2015. *Base Nation: How US Military Bases Abroad Harm America and the World.* New York: Metropolitan Books.

Virilio, Paul. 1989. *War and Cinema: The Logistics of Perception.* London: Verso.

Voigt, Ellen Bryant. 1999. *The Flexible Lyric.* Athens: University of Georgia Press.

Wald, Charles F. 2006. "The Phase Zero Campaign." *Joint Force Quarterly* 43 (October): 72–75.

Wallerstein, Immanuel. 2005. *The Decline of American Power: The U.S. in a Chaotic World.* New York: New Press.

Wedeen, Lisa. 1999. *Ambiguities of Domination: Politics, Rhetoric, and Symbols in Contemporary Syria.* Chicago: University of Chicago Press.

Wehry, Frederic M. 2010. *The Iraq Effect: The Middle East after the Iraq War.* Santa Monica, CA: RAND Corporation.

Weizman, Eyal. 2006. "The Art of War." *Frieze* 99 (May 6). http://frieze.com/article/art-war.

Westheider, James. 2007. *The Vietnam War.* London: Greenwood Press.

Wiest, Andrew. 2013. *Vietnam: A View from the Front Lines.* Oxford: Osprey Publishing.

Williams, Brian Glyn. 2013. *Predators: The CIA's Drone War on Al Qaeda.* Lincoln, NE: Potomac Books.

Williams, Raymond. 1958. *Culture and Society: 1780–1950.* New York: Columbia University Press.

Williams, William Appleman. 2006. *Empire as a Way of Life: An Essay on the Causes and Character of America's Present Predicament, along with a Few Thoughts about an Alternative.* Brooklyn: IG Publishing.

Williams, William Carlos. 1955. "Of Asphodel, That Greeny Flower." *POETRY Magazine* (May).

Witkam, Jan Just. 2007. "Gazing at the Sun: Remarks on the Egyptian Magician Al-Buni and His Work." In *O Ye Gentlemen, Arabic Studies on Science and Literary Culture in Honor of Remke Kruk,* edited by Anand Vrolijik and Jan P. Hogendijk, 183–99. Brill: Boston.

Wolfe, Cary. 2009. *What Is Posthumanism?* Minneapolis: University of Minnesota Press.

Wool, Zoë. 2015. *After War: The Weight of Life at Walter Reed.* Durham, NC: Duke University Press.

Wool, Zoë, and Seth Messinger. 2012. "Labors of Love: Transformation of Care in the Nonmedical Attendant Program at Walter Reed Army Medical Center." *Medical Anthropology Quarterly* 26 (1): 26–48.

Woytowich, Major Adam. 2016. "The Relevance of Robin Sage." *Special Warfare* (July–December).

Wright, C. D. 2007. *One Big Self.* Port Townsend, WA: Copper Canyon Press.

Zani, Leah. 2019a. *Bomb Children: Life in the Former Battlefields of Laos.* Durham, NC: Duke University Press.

———. 2019b. "A Fieldpoem for World Poetry Day." *Anthropology News* 60 (2): e87–e93. https://doi.org/10.1111/AN.1121.

Zia, Ather. 2000. *The Frame: An Anthology of Poems.* Kashmir, India: Cultural Academy of Languages and Arts.

Zinn, Howard, Mike Konopacki, and Paul Buhle. 2008. *A People's History of American Empire.* New York: Metropolitan.

MILITARY AND GOVERNMENT DOCUMENTS

Army Special Operations Forces (ARSOF). 2008. *Unconventional Warfare Field Manual (FM 3-05.130).* Washington, DC: Department of the Army. https://fas.org/irp/doddir/army/fm3-05-130.pdf.

Center for Global Development. 2007. "Phase Zero: The Pentagon's Latest Idea." Center for Global Development Commentary & Analysis, July 20, 2007. www.cgdev.org/blog/phase-zero-pentagons-latest-big-idea.

Congressional Budget Office. 2008. "Contractors' Support of U.S. Operations in Iraq." Prepared by Daniel Frisk and R. Derek Trunkey, under J. Michael Gilmore and Matthew Goldberg. www.cbo.gov/publication/41728.

Department of the Army. 2006. *Field Manual 3-24.* Washington, DC: Headquarters of the Department of the Army.

———. 2007. *U.S. Army Counterinsurgency Handbook.* New York: Skyhorse Publishing.

———. 2008. *U.S. Army Special Forces Handbook.* New York: Skyhorse Publishing.

———. 2014. "Regulation 350-1: Army Training and Leader Development." Washington, DC: Headquarters of the Department of the Army. https://usacac.army.mil/sites/default/files/documents/cace/LREC/AR350-1_Web_FINAL.PDF.

Department of Defense (DOD). 2007a. "Department of Defense Regional and Cultural Capabilities: The Way Ahead." Washington, DC: Department of Defense.

———. 2007b. "Directive No. 5160.70: Management of Language and Regional Expertise." Washington, DC: DOD. http://www.dtic.mil/whs/directives/corres/pdf/516070p.pdf.

———. 2012. "Sustaining U.S. Global Leadership: Priorities for 21st Century Defense." January 2012. Washington, DC: DOD. https://archive.defense.gov/news/defense_strategic_guidance.pdf.

Department of Defense Authorization for Appropriations for Fiscal Year 2007. Hearings Before the Committee of Armed Services, United States Senate, 109th Congress, Second Session on S. 2276.

Finney, Nathan. 2008. *Human Terrain Team Handbook.* Fort Leavenworth, KS. https://info.publicintelligence.net/humanterrainhandbook.pdf.

Government Accountability Office (GAO). 2011. "Defense Language and Culture Training: Opportunities Exist to Improve Visibility and Sustainment of Knowledge and Skills in Army and Marine Corps General Purpose Forces." Washington, DC: U.S. GAO. www.gao.gov/products/gao-12-50.

Joint Chiefs of Staff. 2013. *Joint Intelligence.* Washington, DC: U.S. Army. www.jcs.mil/Portals/36/Documents/Doctrine/pubs/jp2_0.pdf.

SFA (Special Forces Association). n.d. "Inside the Special Forces Qualification Course." Accessed August 27, 2018. www.specialforcesassociation.org/inside-the-sfqc.

U.S. Army John F. Kennedy Special Warfare Center and School. 2011. "News Media Ground Rules."

U.S. House of Representatives Committee on Armed Services. 2008. "Building Language Skills and Cultural Competencies in the Military: DOD's Challenge in Today's Educational Environment." Washington, DC: HASC Oversight & Investigations. http://prhome.defense.gov/Portals/52/Documents/RFM/Readiness/DLNSEO/files/LanguageCultureReport Nov08_HASC.pdf.

U.S. Special Operations Command (USSOCOM). 2020. "Comprehensive Review." Washington, DC: USSOCOM. https://sof.news/pubs/USSOCOM-Comprehensive-Ethics-Review-Report-January-2020.pdf.

REFERENCE WORKS

Almaany Arabic English Dictionary. www.almaany.com/en/dict/ar-en.

Asad, Muhammad. 2012. *The Message of the Qu'ran: The Full Account of the Revealed Arabic Text Accompanied by Parallel Transliteration.* London: The Book Foundation.

Encyclopaedia of Islam, 2nd edition, Brill online. www.brill.com/publications/online-resources/encyclopaedia-islam-online.

Lisan al-Arab Dictionary. https://archive.org/details/lisanalarab07ibnmuoft.

Index

Abu Ghraib, 40, 68, 205
accusation: of betrayal during wartime, 64, 68,
75–80, 83, 146, 192, 244n13, 257n1; of
duplicity (among role-players), 106, 109–
10, 113, 118, 143; of espionage, 90–92;
sectarianism and, 85–87, 228n32; trauma
from, 162–64, 167–68, 214, 259n11
affect: affective surplus, 156–61, 168; as military tactic, 7, 11, 47; subjectivity and,
170, 259n12. *See also* laughscream
Afghanistan: Afghan role-players, 140–43;
American military in, 7, 12–13, 55–56,
77; Gant in, 53; trainings on, 9, 135
'alāsah, 76–80, 104
American Anthropological Association, 39
American Empire, 6, 15, 29, 56. *See* Empire
American military. *See* military, U.S.
Americanness, 102, 107–12, 118
anthropology: colonialism and, 8, 198,
225n16, 239n18, 255n18, 261n12; ethics and, 39; poetry and, 6, 210, 220,
262n3; of war, 8, 14, 30, 198–200,
225n16, 261n12; Writing Culture movement, 210, 216
authenticity: affective, 156–57; military prescriptions of, 126–27, 134–35, 145, 164,
253n4; role-players frustrations with,
139–40

Ba'ath Party, 67, 70, 73. *See* Hussein, Saddam
becoming-imperceptible, 35–36, 52–58, 182
Biden, Joseph, 7, 13, 56
Bush, George W., 39, 56, 68, 104, 246n25

capitalism, 16, 30, 151, 157, 203, 235n5,
246n25
"closed equation of representation" (Deleuze
and Guattari), 130–32, 190
Cold War, 4, 8–9, 182–83, 224n9, 237n10,
253n5
collaboration, 67–69, 79, 85, 113, 192,
258n7. *See also* masking
colonialism: colonial gaze, 46, 51–52; colonial
tropes, 39, 42, 66; history of military, 22
Combat Training Center, 9, 42, 119
conflict ethnography, 38
conscripts of modernity, 69, 233n2
contractors, military: American, 26, 66, 77,
121, 143, 224n6, 238n13, 246n25,
247n27; Iraqi wartime, 4, 7, 62, 69–71,
76, 214; role-players thoughts on, 140,
159, 218; role-playing contracts, 4, 99,
106, 133–36, 156–57, 183, 186, 213,
224n6, 226n19, 248n3
counterinsurgency (COIN): anthropology and,
38–40, 198; logic of, 13, 19, 167, 182;
military history of, 9, 11–12, 38–40;

Founded in 1893,
UNIVERSITY OF CALIFORNIA PRESS
publishes bold, progressive books and journals
on topics in the arts, humanities, social sciences,
and natural sciences—with a focus on social
justice issues—that inspire thought and action
among readers worldwide.

The UC PRESS FOUNDATION
raises funds to uphold the press's vital role
as an independent, nonprofit publisher, and
receives philanthropic support from a wide
range of individuals and institutions—and from
committed readers like you. To learn more, visit
ucpress.edu/supportus.

www.ingramcontent.com/pod-product-compliance
Lightning Source LLC
Chambersburg PA
CBHW020829270326
41928CB00006B/470